Severe and Hazardous Weather

Active Learning Exercises

Robert M. Rauber

John E. Walsh

Donna J. Charlevoix

University of Illinois at Urbana-Champaign

KENDALL/HUNT PUBLISHING COMPANY
4050 Westmark Drive   Dubuque, Iowa 52002

Contents

Preface

These *Active Learning Exercises* that accompany *Severe and Hazardous Weather* were developed for in-classroom use to reinforce concepts taught from the textbook and in lectures. Most exercises can be completed in about ten minutes. In our classes at the University of Illinois, students often complete an exercise right after the material is taught. In this way, students have the opportunity to test their knowledge, and ask immediate follow-up questions to clarify points they might not have understood in lecture. The exercises can also be used as homework assignments to reinforce concepts presented in class or encountered in the textbook.

Enjoy *Severe and Hazardous Weather*!

Bob Rauber
r-rauber@uiuc.edu

John Walsh
walsh@atmos.uiuc.edu

Donna Charlevoix
charlevo@atmos.uiuc.edu

Name: _____

Section: _____ Date: _____

Exercise 1.1 Geography Overview

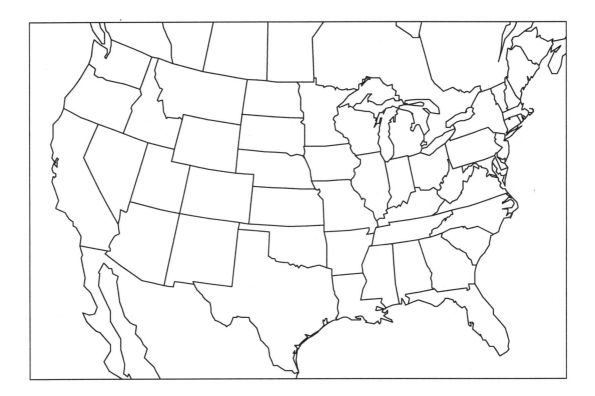

Imagine hearing about a hazardous weather event with no reference to the location where it happened! Or worse yet, hearing the location and not knowing where it is! Students of hazardous weather must have a basic knowledge of geography. Test your knowledge of geography by completing the following exercise.

On the map above:

1. Draw a line along the crest of the Rocky Mountains and label it "Rockies."

2. Draw a second line along the crest of the Sierra Nevada and label it "Sierras."

3. Draw a third line along the Cascade Range and label it "Cascades."

4. Draw a fourth line along the Appalachian Mountains and label it "Appalachians."

5. Label the Atlantic Coastal Plain and the Great Plains.

6. Label each of the Great Lakes.

7. Label the Pacific Ocean, the Atlantic Ocean, and the Gulf of Mexico.

8. Identify the following states: Pennsylvania (PA), Michigan (MI), North Carolina (NC), Colorado (CO), Nebraska (NE), Oregon (OR), and Alabama (AL).

Name: _____

Section: _____ Date: _____

Exercise 1.2 Temperature, Pressure and Moisture

1. Convert the following temperatures from either Fahrenheit to Celsius or Celsius to Fahrenheit.

    32°F = _____   30°C = _____

    68°F = _____   10°C = _____

    86°F = _____   -10°C = _____

2. At approximately what altitudes (in kilometers) are the following pressures observed?

    1013 mb   _____

    850 mb   _____

    500 mb   _____

    300 mb   _____

3. The following observations were taken at three cities:

    Pittsburgh, PA   Temp = 23°F   Dewpoint Temp = 21°F

    Tampa, FL   Temp = 78°F   Dewpoint Temp = 53°F

    Phoenix, AZ   Temp = 100°F   Dewpoint Temp = 35°F

4. Which city has the highest vapor pressure?   _____

5. Which city has the highest saturation vapor pressure? _____

6. Which city has the highest relative humidity?   _____

Exercise 1.3 Latent Heat

Phase changes of water (conversions between ice, liquid water, and water vapor) occur constantly in Earth's atmosphere. Heat is either released into the atmosphere (heating the air) or extracted from the atmosphere (cooling the air) during phase changes. In the exercises below insert the appropriate letters to indicate the correct phase change and whether the atmosphere would be heated or cooled as a result of the phase change.

V = vapor   V = vapor   H = heating
L = liquid   L = liquid   C = cooling
I = Ice   I = Ice

1.   Evaporation: _____ converts to _____ which leads to _____ of the air.

2.   Melting   _____ converts to _____ which leads to _____ of the air.

3.   Condensation _____ converts to _____ which leads to _____ of the air.

4.   Deposition   _____ converts to _____ which leads to _____ of the air.

5.   Sublimation   _____ converts to _____ which leads to _____ of the air.

6.   Freezing   _____ converts to _____ which leads to _____ of the air.

7.   Which of the six processes listed above releases the greatest amount of heat to the atmosphere?

8.   Which of the six processes extracts the most heat from the atmosphere?

9.   What types of particles are clouds composed of?

10. During cloud formation, do changes of phase warm or cool the atmosphere? Why?

Name: _____

Section: _____ Date: _____

Exercise 1.4 Seasonal Temperature Variations

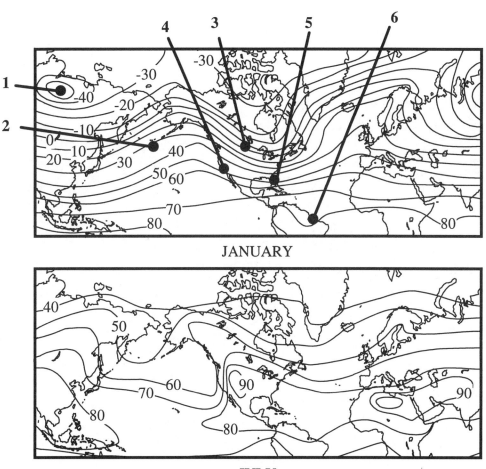

JANUARY

JULY

From the maps of average January and July global temperatures, estimate the average seasonal temperature change in degrees Fahrenheit at the following locations:

1. Northern Siberia _____   2. The Aleutian Islands _____

3. Regina, Canada _____   4. Los Angeles, CA   _____

5. Miami, Florida _____   6. Amazon River Delta _____

7. Over the globe, where are the smallest seasonal temperature changes found?

8. Where are large seasonal temperature changes found?

Name: _____

Section: _____   Date: _____

Exercise 2.1 Time Conversions

Fill in the blanks in the table below to complete the conversions between Universal Coordinated Time (UTC) and Standard Time or Daylight Time in the indicated time zones of the United States: Eastern (EST, EDT), Central (CST, CDT), Mountain (MST, MDT) or Pacific (PST, PDT).

1. 4 p.m.   EST   Tuesday   _____   UTC   _____

2. 7 a.m.   PDT   Monday   _____   UTC   _____

3. 10 p.m. MDT   Thursday   _____   UTC   _____

4. _____   CDT   _____   1500   UTC   Wednesday

5. _____   EDT   _____   0000   UTC   Friday

6. _____   MST   _____   1200   UTC   Saturday

7. 2200   PDT   Thursday   _____   UTC   _____

8. 0300   CDT   Tuesday   _____   UTC   _____

9. 1800   MST   Saturday   _____   UTC   _____

10. _____   CST   _____   0200   UTC   Sunday

Name: _____

Section: _____ Date: _____

Exercise 2.2 Interpreting a Sounding

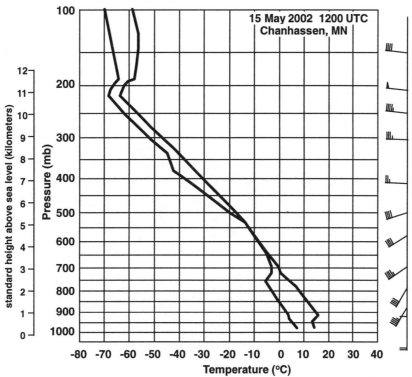

Shown above is a sounding for Chanhassen, MN. Use the plotted information to answer the following questions.

1. What is the surface air temperature? _____

2. What is the surface pressure? _____

3. What is the surface dewpoint? _____

4. What is the 850 mb dewpoint depression? _____

5. At what height is the tropopause? _____

6. What is the strongest wind above Chanhassen? _____

7. Which layer is a cloud layer above Chanhassen? (Identify the layer by the pressures at the top and bottom). _____

8. What is the wind speed and direction at 500 mb? _____

9. Which layer is an inversion layer in the lower atmosphere? _____

Exercise 2.3 Instruments Used for Weather Observations

Indicate at least one instrument that can provide the following weather information. (In some cases, there may be more than one correct answer.)

1. The 500 mb temperature is –6 °C. _____.

2. The sea level pressure is 1021.6 mb. _____

3. A thunderstorm is approaching. _____.

4. Precipitation during the past 6 hours has been 0.3 inches. _____.

5. The wind at 300 mb is 120 knots. _____.

6. A hurricane is 400 miles offshore of the Gulf Coast. _____.

7. Freezing rain is falling. _____.

8. The dewpoint is 53°F at the surface. _____.

9. The height of the 850 mb pressure is 1,530 meters. _____.

10. The wind 3 km above the ground has changed direction in the last hour. _____.

11. The dewpoint depression at 700 mb is 14°C. _____.

12. The temperature of the top of the highest cloud layer is –52 °C. _____.

Name: _____

Section: _____ Date: _____

Exercise 2.4 Satellite and Radar Observations

Listed below are types of information that can be obtained from satellites and radar. In each case, match the product with letter denoting the source of information.

Source of information:

A.	Radar reflectivity	D.	Satellite visible image
B.	Radar radial velocity	E.	Satellite infrared image
C.	Radar wind profiler	F.	Satellite water vapor image

Observational information:

1. _____ Rain is falling at a rate of 3 cm/hr at your location.

2. _____ Lake Michigan is warmer than the surrounding land area.

3. _____ Dry air in the middle troposphere extends from Idaho to Michigan in a band approximately 300 miles wide.

4. _____ The cloud tops on a line of thunderstorms approaching your location reach 45,000 feet.

5. _____ Your location is snow-free, but snow covers the ground 150 miles to the north.

6. _____ There is rotation in a thunderstorm approximately 30 miles to the south of your location.

7. _____ The winds at 5 km are northwest at 55 knots.

8. _____ Low stratus clouds are present in the area to your west.

9. _____ A line of dissipating cumulonimbus clouds without precipitation is approaching your location.

10. _____ Precipitation in a thunderstorm to your west is decreasing in intensity.

11. _____ A band of altostratus clouds is approaching your location at midnight.

12. _____ More than 3 inches of rain fell at your location in the past 12 hours.

Exercise 3.1 Decoding Observations

1. Decode the following surface station and upper air station observations. Be sure to include units.

surface station

temperature   _____

dewpoint temperature _____

cloud cover   _____

wind direction _____

wind speed   _____

pressure   _____

significant weather   _____

upper air station (500 mb)

temperature   _____

dewpoint depression   _____

dewpoint temperature _____

wind direction _____

wind speed   _____

height _____

height change _____

2. Create station models for the following data.

surface

temperature = 88°F
dewpoint = 85°F
wind direction = southeast
wind speed = 20 kts
pressure = 1002.4 mb
significant weather = thunderstorm
cloud cover = 100%

upper air (200 mb)

temperature = −61°C
dewpoint temperature = −57°C
wind direction = west
wind speed = 75 kts
height = 12,210 m
height change = −100 m

Name: _____

Section: _____   Date: _____

Exercise 3.2 Contouring

Rules for Contouring

- Contour lines are drawn to identify constant values of an atmospheric variable.
- A contour is drawn through the station location only if the data for that station has the exact value of the contour; otherwise the contour is drawn between stations.
- Higher values are on one side of the contour and lower values on the other side.
- Contours are drawn at equal increments of the contoured variable.
- Contours never cross or touch each other.
- More than one contour of a given value may appear on a given map.
- All contour lines must be clearly labeled.
- Contours always form closed loops on world maps.
- Isotherms (temperature) and Isodrosotherms (dewpoint temperature) are typically drawn using 5°F or 10°F (or 5°C) intervals, isobars (pressure) at 4 mb intervals.

1. The diagram below shows the numbers 29, 30 & 31. Draw a contour for the number "30."

2. The map of the Midwest shows temperatures in degrees Fahrenheit. Draw contours every 5°F. Label your contours.

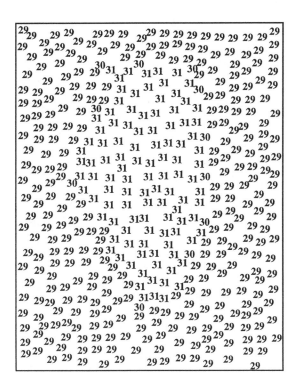

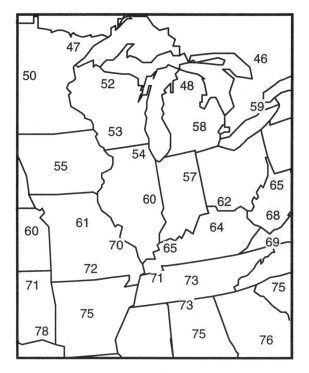

Name: _____

Section: _____ Date: _____

Exercise 3.3 Constant Pressure Charts

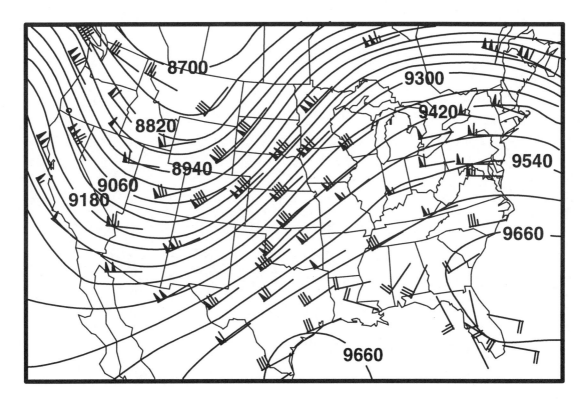

The map above depicts the conditions at 300 mb.

1. Draw a solid line to show the axis of the large trough over the western United States.

2. Draw a dashed line to show the axis of the large ridge over the eastern United States.

3. Draw the 100 knot isotachs (lines of constant wind speed).

4. Draw a line along the axis of the 300 mb jetstream.

5. Place an "H" in the two locations where the 300 mb surface is at its highest altitude.

6. Place an "L" in the location where the 300 mb surface is at its lowest altitude.

7. Based on this map, what can you say about the relationship between the direction and speed of the wind and the orientation and spacing of the height contours?

Exercise 3.4 Slope of Pressure Surfaces

Plot the altitude of each pressure surface on the vertical line above each location. Connect the points for each pressure surface to visualize the slope. Plot the altitude of the phenomena listed in the second table on the vertical line to the right of the diagram.

	Approximate Altitude (m)		
	Southern Texas	Nebraska	North Dakota
850 mb	1,540	1,500	1,480
700 mb	3,180	3,000	2,140
500 mb	5,800	5,500	5,000
300 mb	9,800	9,600	9,400
250 mb	11,000	10,500	10,300
200 mb	12,500	12,000	11,750

Mt. Everest	8,800 m	Sketch the mountain
Commercial aircraft flight	10,500 m	Draw an airplane
Base of cumulus cloud	1,000 m	Sketch a cloud

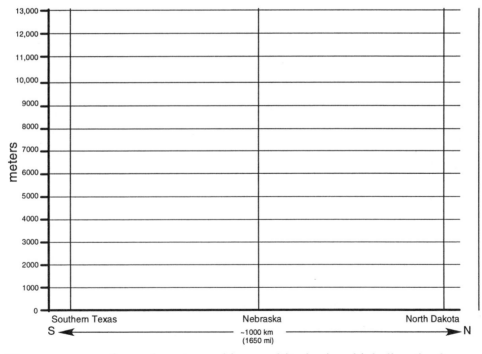

1. The pressure surfaces slope toward lower altitudes in which direction?

2. Based on the slope of the pressure surfaces, would the temperature in North Dakota be warmer or colder than the temperature in Texas? How did you determine this?

Name: _____

Section: _____   Date: _____

Exercise 4.1 Resolution of Numerical Model Grids

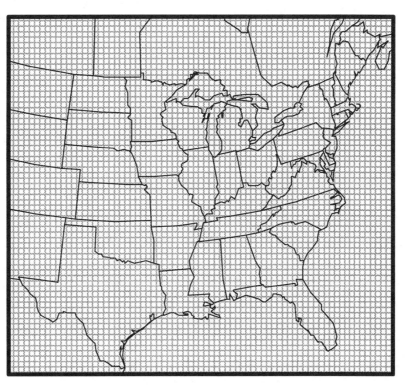

A weather phenomenon is well-resolved by a model if there are 8 or more gridpoints across its shortest dimension, resolved if there are between five and 7 points, poorly resolved if there are 2-4 points, and unresolved if there are less than 2 points. The grid above is similar to that used in numerical models. The spacing of the points is approximately 50 km. On the grid, draw a box that estimates the size of each weather phenomenon listed below. In the blanks, write whether the phenomenon is well-resolved (W), resolved (R), poorly resolved (P), or unresolved (U) on this grid.

Phenomenon	*Size*	*Resolvable?*	*# grid points*
1. hurricane	(300 km × 300 km)	_____	_____
2. thunderstorm	(20 km × 20 km)	_____	_____
3. cold frontal cloud band	(200 km × 1000 km)	_____	_____
4. sea breeze	(20 km × 200 km)	_____	_____
5. tornado	(200 m × 200 m)	_____	_____
6. extratropical cyclone	(1500 km × 1500 km)	_____	_____
7. freezing rain band	(50 km × 300 km)	_____	_____
8. lake-effect snow band	(20 km × 200 km)	_____	_____

13

Name: _____

Section: _____ Date: _____

Exercise 4.2 Initialization of a Numerical Model

Station reports from five locations in the United States are shown below. These reports were among the observations used to initialize the Eta forecast model. The initialized field is shown below the surface reports.

1. Decode the sea level pressure for each station and enter the value in the table below.
2. Use the map to determine the pressure in the Eta model's initialization at each station's location, and record these values in the table.
3. compute the difference between the reported pressure and model's initialization ($P_{report} - P_{initialization}$). Include the units.

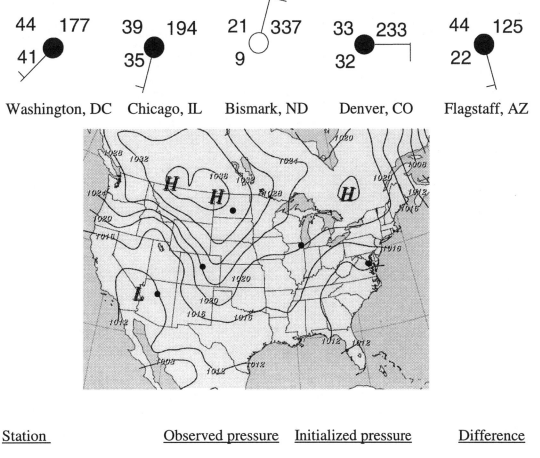

Station	Observed pressure	Initialized pressure	Difference
Washington, D.C.	_____	_____	_____
Chicago, Illinois	_____	_____	_____
Bismarck, North Dakota	_____	_____	_____
Denver, Colorado	_____	_____	_____
Flagstaff, Arizona	_____	_____	_____

14

Name: _____

Section: _____   Date: _____

Exercise 4.3 Accuracy of a Numerical Model Forecast

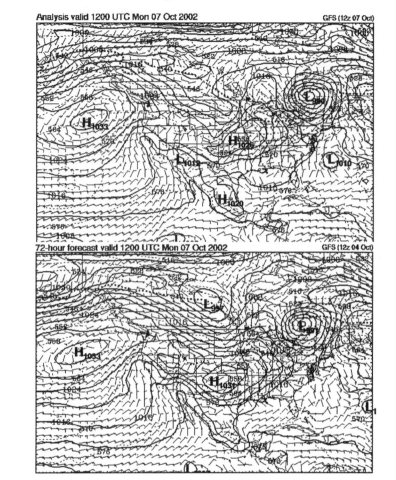

The top map is the analysis of sea level pressure (solid dark lines) and winds for the GFS model for 1200 UTC 7 October 2002. The bottom map is a forecast of sea level pressure and winds made 72 hours earlier that is valid for the same time. Evaluate the forecast by examining errors in the location and intensity (central pressure) of the following features:

	Location correct (y/n)?	*Pressure error ($P_{anal} - P_{forecast}$)*
High pressure system over the central United States:	_____	_____ mb
High pressure system over the eastern Pacific Ocean:	_____	_____ mb
Low pressure system northeast of the Great Lakes:	_____	_____ mb
Low pressure system over northern Canada:	_____	_____ mb

15

Name: _____

Section: _____   Date: _____

Exercise 4.4 Model Resolution and Topography

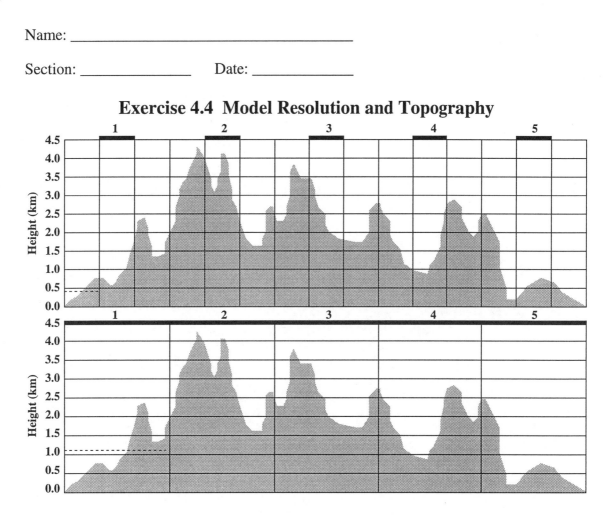

The diagrams above show the same mountain range with horizontal grids superimposed. The top grid has three times the horizontal resolution of the bottom grid.

1. Draw horizontal lines to show the "average" mountain height in each grid column. The first columns are done for you (see dashed lines).

2. If air was flowing from left to right, on which grid would air have to rise the most to transit across the mountains?

3. The amount of snowfall that falls in mountainous areas during winter storms is closely related to the distance that air must be lifted to cross the mountains. Discuss briefly the effects "gridding" topography might have on errors in precipitation forecasts.

Name: _____

Section: _____   Date: _____

Exercise 5.1 Environmental Lapse Rates

The sounding below shows the temperature measured over a single location. Compute the environmental lapse rate for each layer of the atmosphere listed below. Use the standard atmosphere altitudes on the bar on the left to determine altitudes.

surface to 1 km   _____

1 to 2 km   _____

2 to 5 km   _____

5 to 7 km   _____

7 to 9 km   _____

9 to 10 km   _____

10 to 11 km   _____

11 to 12 km   _____

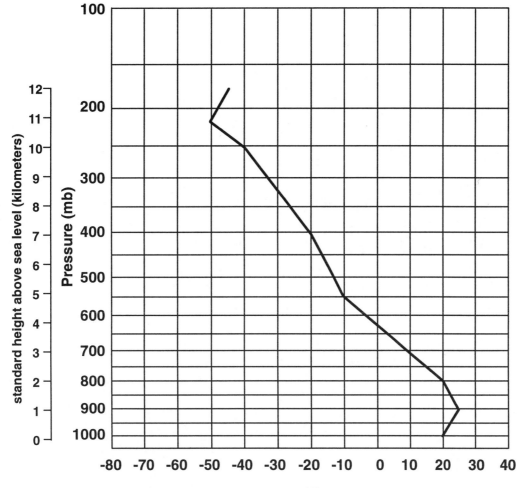

Name: _____

Section: _____   Date: _____

Exercise 5.2 Stability

The stability of the atmosphere helps to determine the types of clouds and severe weather that may develop, if any. The three categories of atmospheric stability are: (1) absolutely stable, (2) absolutely unstable, and (3) conditionally unstable. To determine the stability of the atmosphere you need to know:

1. The Environmental Lapse Rate (ELR);
2. If the air parcel that is lifted is saturated (RH=100%) or unsaturated (RH < 100%).

Other useful information:

  DALR = dry adiabatic lapse rate = 10°C/km (use when RH < 100%)
  MALR = moist adiabatic lapse rate ~ 6°C/km (use when RH = 100%)

---

1. Assume that the air is completely dry. The ELR in the lowest 1 km is 4°C/km. What is the stability of this atmosphere: (1) absolutely stable, (2) absolutely unstable, or (3) conditionally unstable? Why?

2. Assume that the air is saturated. The ELR in the lowest 1 km is 12°C/km. What is the stability of this atmosphere: (1) absolutely stable, (2) absolutely unstable, or (3) conditionally unstable? Why?

3. The ELR in the lowest 1 km is 8°C/km. What is the stability of this atmosphere? (1) stable, (2) conditionally unstable, or (3) absolutely unstable? (Hint: To do this problem, first test a dry parcel and then test a saturated parcel. If the atmosphere is stable with respect to the dry parcel but unstable with respect to the saturated parcel, then the atmosphere is conditionally unstable.)

Name: _____

Section: _____ Date: _____

Exercise 5.3 Stability and Soundings

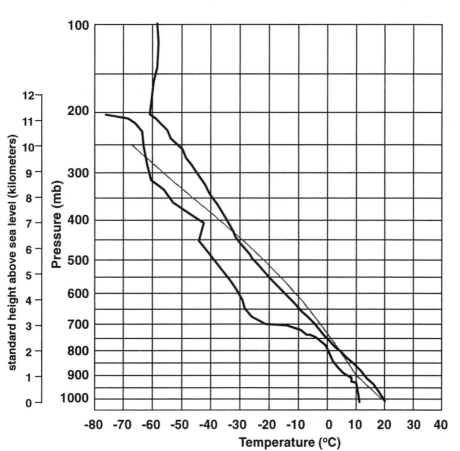

On the sounding above, the temperature and dewpoint temperature are plotted as thick black lines. The temperature of a lifted parcel is also shown as a gray line. Use the sounding to answer the following questions.

1. Which thick line is the temperature, the right line or the left line? _____

2. At what pressure level is the level of free convection? _____

3. If a parcel is lifted to the level of free convection, to what pressure level will buoyancy enable it to rise?

4. At what pressure level is the lifting condensation level? _____

5. At what pressure level is the cloud base? _____

6. Estimate the value of the Lifted Index. _____

Name: _____

Section: _____ Date: _____

Exercise 5.4 Lifted Index and Thunderstorm Development

The map below shows the value of the Lifted Index (contoured) and the locations of fronts on a day in late spring. Based on the information on the map, determine where thunderstorms might form. Assume that there is sufficient moisture available. Place the thunderstorm symbol in the location or locations where thunderstorms are most likely to develop. Below the map, explain in a few words why you picked the location you did.

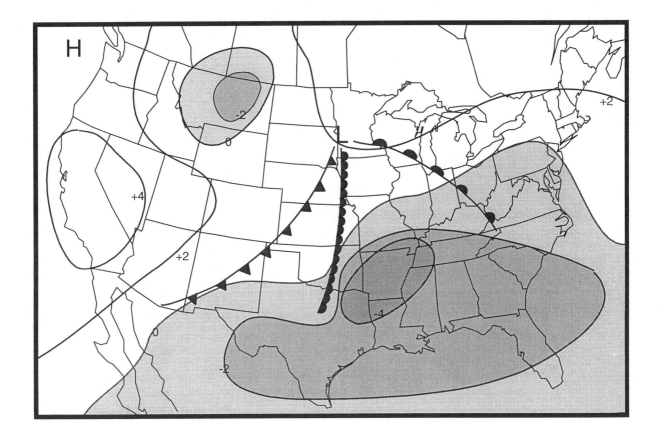

Name: _____

Section: _____ Date: _____

Exercise 6.1 Forces and Force Balances

HPGF = Horizontal Pressure Gradient Force
VPGF = Vertical Pressure Gradient Force
CF   = Coriolis Force
FR   = Frictional Force
GR   = Gravitational Force
GB   = Geostrophic Balance
HB   = Hydrostatic Balance

Place letters on the blanks in each statement to make the statement correct.

1. In the Northern Hemisphere, the _____ always acts opposite the direction of air

   motion, but the _____ always acts to the right of the air motion.

2. Air neither falls to the ground nor drifts into space because the atmosphere is in

   _____, a balance between the _____ and the _____.

3. The _____ is zero at the equator and increases with latitude.

4. The _____ acts to accelerate air horizontally from rest.

5. The _____ can change the direction air moves, but not its speed.

6. If the earth did not rotate, the _____ would be zero and air in motion could

   never be in _____.

7. The _____ is strongest near the earth's surface and decreases in importance at

   higher altitudes.

Exercise 6.2 The Horizontal Pressure Gradient Force

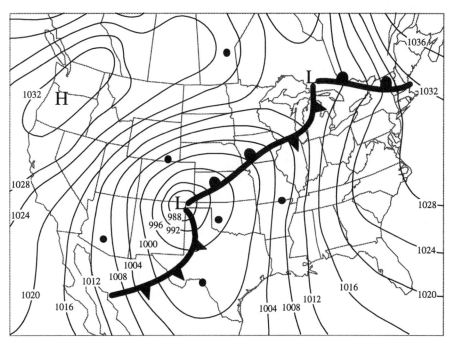

1. At each of the large black dots on the map above, draw an arrow emerging from the dot to indicate the direction of the pressure gradient force. Adjust the length of your arrows so that a long arrow indicates a relatively strong pressure gradient and a short arrow indicates a relatively weak pressure gradient force.

2. What is the average value of the pressure gradient force between the high pressure center on the West Coast (1033 mb) and the low pressure center in Colorado (888 mb)? Assume the distance between these two points is approximately 1500 km.

3. Salt Lake City, in northern Utah about halfway between the high and low, is approximately 20 kilometers in diameter. What is the approximate pressure change across Salt Lake City?

Name: _____

Section: _____ Date: _____

Exercise 6.3 The Geostrophic Wind

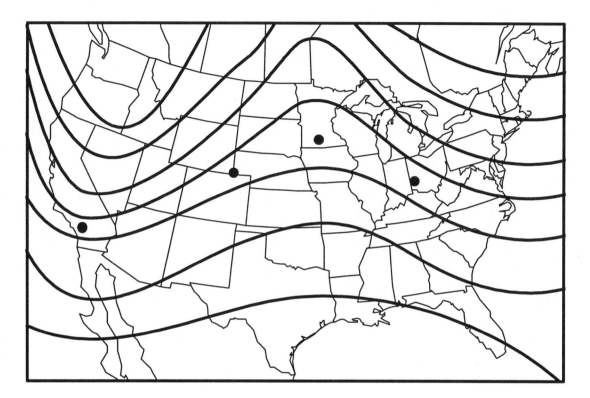

Assume that the flow represented on the 500 mb map above is in geostrophic balance.

1. Draw an arrow showing the speed and direction of the geostrophic wind at each of the four points indicated on the map. The "tail" of each arrow should be at the point and the length of the arrow should represent the relative wind speed (i.e., the point with the strongest wind should have the longest arrow and the point with the weakest wind should have the shortest arrow).

2. Draw a second arrow at each point that represents the pressure gradient force (PGF). Again, the tail of the arrow should be at each point, the arrow should point in the direction the force acts, and the length of the arrow should represent the relative strength of the PGF.

3. Do the same as in (2), but for the Coriolis Force.

Name: _____

Section: _____   Date: _____

Exercise 6.4 Fronts and the Jetstream

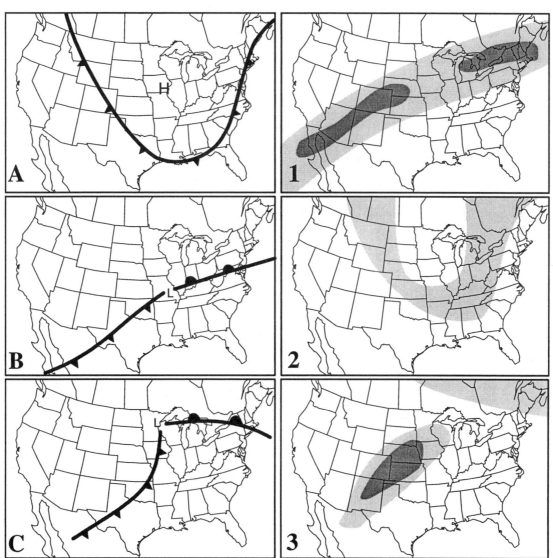

1. The maps on the left side above are each surface maps depicting positions of frontal boundaries and low and high pressure centers. The maps on the right side depict the wind speed at 300 mb, with the light shading indicating winds exceeding 70 knots and the dark shading indicating winds exceeding 100 knots. Each map on the left matches one of the three maps on the right. Which matches which?

Map A matches Map _____   Map B matches Map _____
Map C matches Map _____

2. How did you determine the matching?

Name: _____

Section: _____   Date: _____

Exercise 7.1 Jetstreaks and Curvature

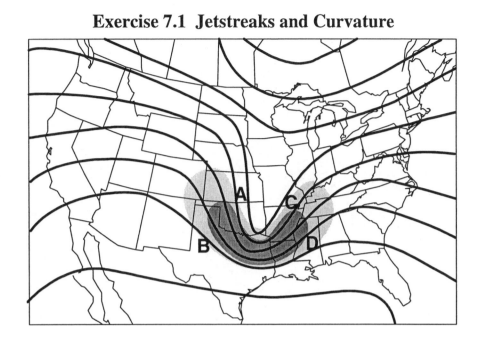

The solid lines on figure above show the 300 mb height field over the United States. Winds within the lightly shaded region exceed 100 knots, and within the dark shaded region exceed 140 knots. Answer the following questions by placing the letters A, B, C or D in the blanks.

1. The right exit region of the jetstreak is located at _____.

2. The left entrance region of the jetstreak is located at _____.

3. Divergence due to flow curvature is occurring at _____ and _____.

4. Convergence due to the jetstreak effect is occurring at _____ and _____.

5. Convergence due to flow curvature is occurring at _____ and _____.

6. Divergence due to the jetstreak effect is occurring at _____ and _____.

7. The maximum convergence is occurring at _____.

8. The maximum divergence is occurring at _____.

9. The surface pressure is rising most rapidly at _____.

10. The surface pressure is falling most rapidly at _____.

Name: _____

Section: _____ Date: _____

Exercise 7.2 Effects of Convergence and Divergence on Surface Systems

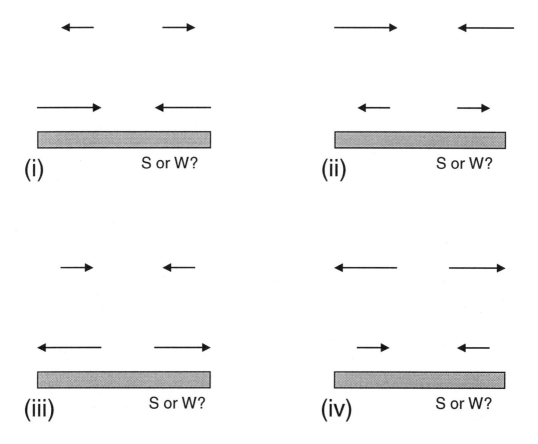

Use the schematic cross-sectional diagrams above to answer (a) through (c). The diagrams indicate convergence or divergence near the surface and in the upper troposphere above surface pressure centers (Highs or Lows). In each case, the length of the arrows indicates the strength of the convergence or divergence. For each diagram:

1. Write "H" or "L" between the arrows at the surface to show whether the **surface pressure center** is a High or a Low.

2. Circle "S" or "W" to indicate whether the surface system will **S**trengthen or **W**eaken.

3. Draw an arrow **above each surface pressure center** to indicate the direction of vertical air motion (upward or downward).

Exercise 7.3 Friction and Its Effect on Winds

In each of the following questions, circle the correct answer from the choices in parentheses.

1.   Friction (increases, decreases) the wind speed.

2.   Friction results in a reduction of the (pressure gradient force, Coriolis force).

3.   Which upper-air pressure is generally closer to the top of the friction layer? (850 mb, 300 mb)

4.   Friction ultimately causes the wind to deflect (into, out of) a surface high-pressure system.

5.   Friction causes (convergence, divergence) of surface winds around a low-pressure center.

6.   Friction contributes to (upward, downward) motion above a surface high-pressure center.

7.   The friction layer will generally be deeper at (mid-afternoon, sunrise).

8.   In airflow that would otherwise be in geostrophic balance, friction causes a (rightward, leftward) deflection.

9.   Friction causes air to spiral (inward, outward) around a surface low-pressure center.

10.   In the Southern Hemisphere, the Coriolis force acts 90 degrees to the left of the wind. In this case, friction will act (in the same direction as, opposite to) the wind.

11.   Over water, friction deflects the wind from its geostrophic direction by an angle that is typically (10°-20°, 20°-40°).

12.   The angle by which friction deflects the wind from its geostrophic value generally becomes (larger, smaller) with increasing altitude in the atmosphere.

13.   If the only convergence and divergence in the atmosphere resulted from friction, surface highs would (strengthen, weaken) over time, while surface lows would (strengthen, weaken) over time.

Name: _____

Section: _____ Date: _____

Exercise 7.4 High and Low Pressure Centers

Listed below are characteristics of high- or low-pressure centers in the Northern Hemisphere. In each case, indicate by **H** or **L** whether the statement applies to high- or low-pressure centers. If a statement is valid for both types of centers, place both **H** and **L** in the blank space.

1. _____ Winds blow counter-clockwise.

2. _____ Winds spiral inward toward the pressure center.

3. _____ Vertical motions are downward above the surface pressure center.

4. _____ Clouds and precipitation generally occur above the surface pressure

   center.

5. _____ Associated with convergence in the upper troposphere.

6. _____ Surface winds are deflected to the left of the isobars.

7. _____ Generally located to the east of an upper-air trough.

8. _____ Surface winds are divergent.

9. _____ Favored by cooling of an air column.

10. _____ Pressure gradient force points away from the pressure center.

11. _____ Release of latent heat favors intensification of the surface pressure center.

12. _____ Lower pressure is to the left of the winds.

13. _____ Associated with subsidence, dry air and generally clear skies.

14. _____ Wind speeds (in the absence of friction) are slower than geostrophic.

Name: _____

Section: _____   Date: _____

Exercise 8.1 Airmass Identification

For each of the following reports of temperature and dewpoint, identify the type of airmass that is affecting the city. Use the symbols m or c for maritime or continental followed by P or T for Polar or Tropical (for example: mP). In the final column, indicate the wind direction (N, NE, E, SE, S, SW, W, NW) that is likely to have brought the airmass to the city.

	Month	City	Temp. (°F)	Dewpt. (°F)	Airmass	Wind direction
1.	April	Boston, MA	45	43		
2.	Feb.	Boston, MA	32	12		
3.	June	Atlanta, GA	79	73		
4.	Jan.	Wichita, KS	16	2		
5.	June	Wichita, KS	95	49		
6.	June	Kansas City, MO	93	74		
7.	Oct.	Seattle, WA	45	44		
8.	Mar.	Seattle. WA	40	21		
9.	Feb.	Denver, CO	28	27		
10.	June	Minneapolis, MN	74	47		
11.	Sept.	Bismarck, ND	68	38		
12.	Dec.	Louisville, KY	54	51		
13.	Sept.	San Francisco, CA	88	48		

Name: _____

Section: _____   Date: _____

Exercise 8.2 Which Front Passed the Station?

Each pair of station models below shows data from the same station. However, the data were taken six hours apart and a front passed the station between the two times. Based on the station models, determine what type of front passed the station. Choose from the following list and use each selection only once:

C   Cold Front   D   Dry Line
W   Warm Front   U   Upper Level Front
O   Occluded Front

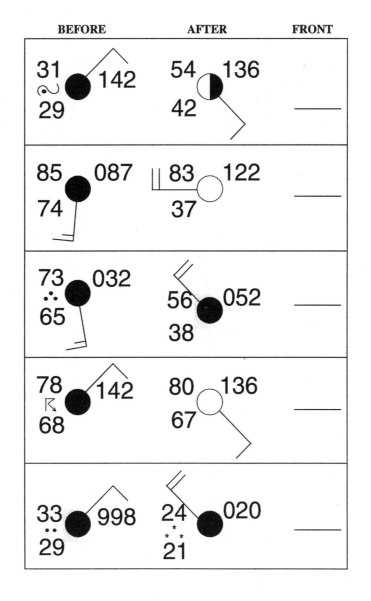

Name: _____

Section: _____ Date: _____

Exercise 8.3 Where Is the Front?

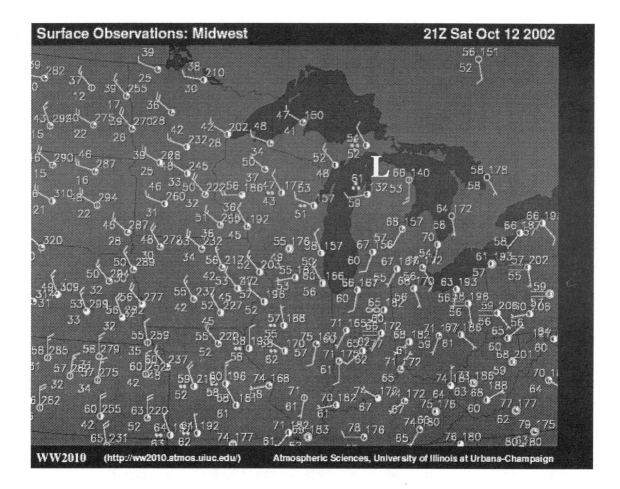

Carefully examine the station data on this surface map for 2100 UTC, 12 October 2002.

Can you find the cold front?

1. Draw in the cold front with the correct frontal symbols.

2. How did you determine the location of the front? (e.g. what data did you use?)

Name: _____

Section: _____   Date: _____

Exercise 8.4 Vertical Structure of a Front

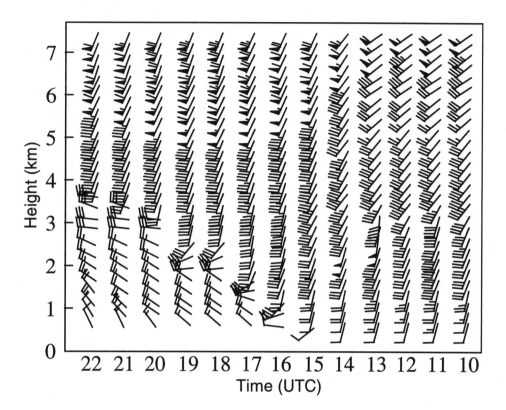

Shown above is a ten-hour time sequence of data from a wind profiler during the passage of a cold front. Answer the following questions based on the wind profiles obtained from the wind profiler. (Note: Wind profiler data is plotted with the earliest time on the right and the latest on the left, which has the effect of making changes in time look the way they appear in space.)

1. How can this data be used to identify a front? _____

2. What time did the front arrive at the station? _____

3. Is there a low level jet ahead of this front? _____

4. At what level and time did the center of the low level jet pass over the station?
    _____

5. At what level and time did the center of the higher level jetstream pass the station? _____

6. How deep is the cold airmass over the station at 22 UTC? _____

32

Name: _____

Section: _____ Date: _____

Exercise 9.1 Weather Conditions within a Cyclone

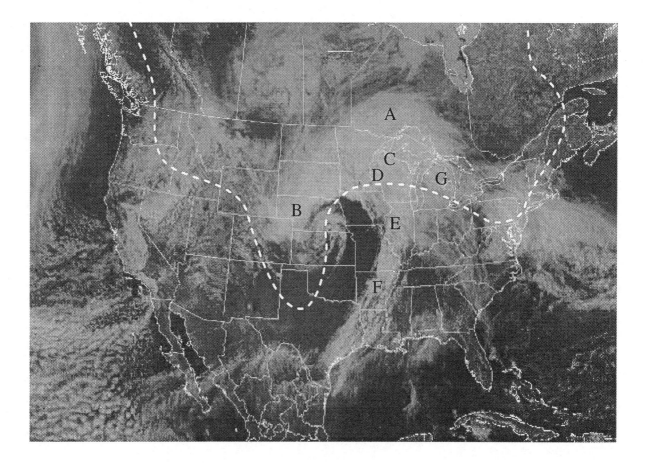

Shown above is a visible satellite image of a cyclone over the central United States. The dashed line is the 0°C isotherm at the surface. Assume that the cyclone occurred during the first week of March.

1. Place an L at the location of the low pressure center.

2. Draw a line along the leading front south of the low pressure center.

3. Draw station models that show temperature, wind speed and direction, cloud cover, and weather (use standard weather symbols) near the 7 points labeled A-G on the image.

Exercise 9.2 Jetstreaks, Troughs and Surface Low-Pressure Centers

Shown below are four 300 mb maps covering a 36 hour period during which a strong cyclone developed and moved across the Great Plains. On each map, place an "L" at the most likely position of the <u>surface</u> low-pressure center based on the position and orientation of the jetstream. Sketch a cold front, an upper level front, and a warm front in a reasonable position on each diagram.

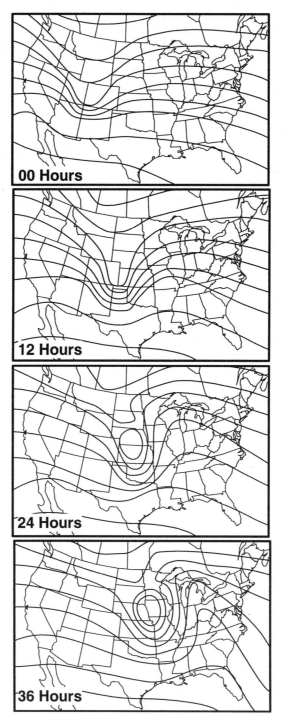

00 Hours

12 Hours

24 Hours

36 Hours

Name: _____

Section: _____   Date: _____

Exercise 9.3 Soundings through Cyclones

The map to the left shows a cyclone over the Great Plains.

The soundings at the bottom of the page were taken at each of the six points labeled A through F on the map. Write a letter in the blank on each of the soundings to indicate which sounding corresponds to each point A-F.

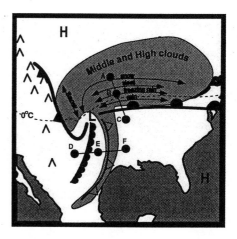

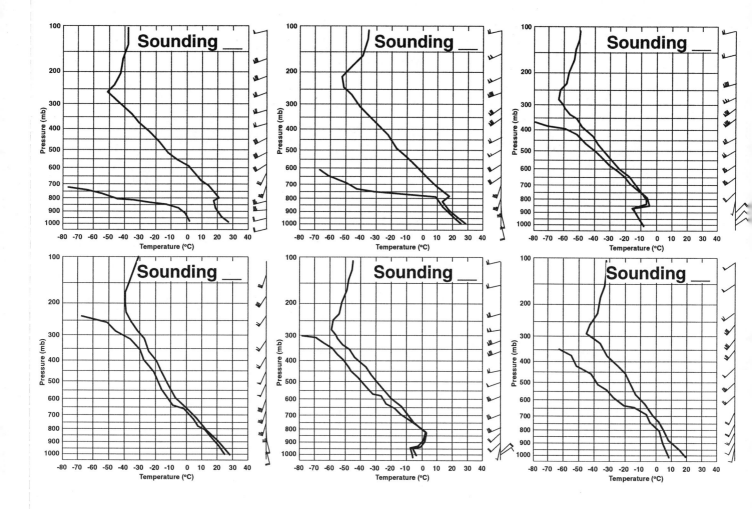

Name: _____

Section: _____   Date: _____

Exercise 9.4 Vertical Structure of Fronts in a Cyclone's Southern Sector

The left panels of the figure below show three cyclones over the Great Plains, each with different frontal structure south of the low-pressure center. The right panels are cross sections along the direction of the arrow in the corresponding figure to the left. The topography and a thunderstorm are shown on each cross section.

On the cross section diagram, sketch the fronts as they would appear in the vertical.

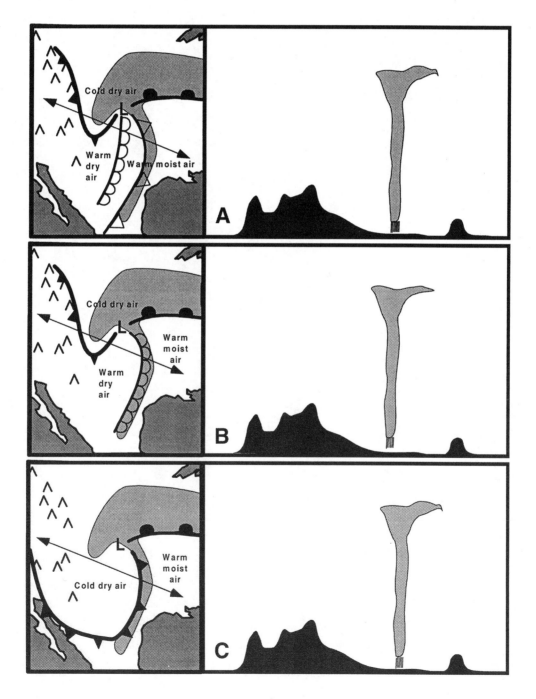

Name: _____

Section: _____ Date: _____

Exercise 10.1 The Rain-Snow Line and Nor'easters

Shown below is a map of New York City area with the city itself shaded lightly, and the metro-area shaded darker. The boxes are 10 kilometers on a side. Three Nor'easters pass up the coast during the winter season. The list below shows, for each Nor'easter, 1) the temperature of air over the ocean, 2) the rate that air cooled after it first moved over a land surface, and 3) the wind speed in kilometers/hour. In all three cases, the wind was directly from the east.

Nor'easters:

1. *Temperature of air over ocean* = 44°F, *Cooling rate* = 2°F/hr, *Wind* = 20 km/hr

2. *Temperature of air over ocean* = 38°F, *Cooling rate* = 3°F/hr, *Wind* = 20 km/hr

3. *Temperature of air over ocean* = 33°F, *Cooling rate* = 3°F/hr, *Wind* = 20 km/hr

(a) Draw in the rain-snow line at the time the observations above were taken for each of the three Nor'easters listed above.

(b) If a forecaster issued a forecast for heavy snow in New York, would the forecast be a boom or a bust for each case? How about if the forecast was issued for the Metro area rather than the city itself?

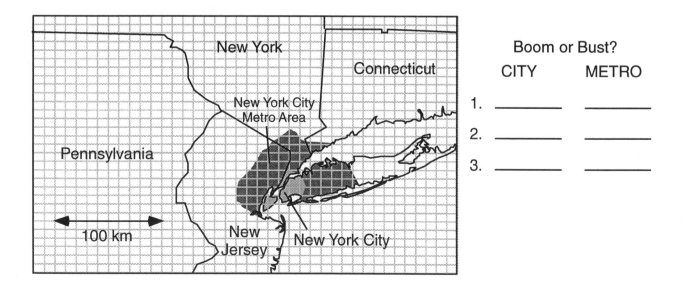

37

Name: _____

Section: _____ Date: _____

Exercise 10.2 Interacting Jetstreak Circulations

Below are two maps showing the height field (solid lines) and jetstreak locations where the wind speed exceeds 100 knots (shaded).

1. On both maps, place a "D" in the quadrants of each jetstreak where divergence is occurring. Place a "C" in the quadrants of each jetstreak where convergence is occurring.

2. On Map B, place an "L" where an intense low pressure system is likely to develop at the surface. Explain briefly below why you chose that location.

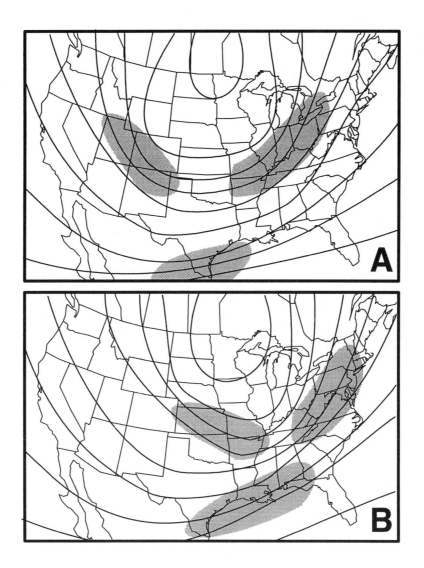

Name: _____

Section: _____ Date: _____

Exercise 10.3 Where Is the Low-Pressure Center?

1. On each map below, place an "L" at the location where a strong surface low pressure center would be located to produce a heavy snowfall in the highlighted city.

2. Sketch in the probable location of the cold front and warm front for each map.

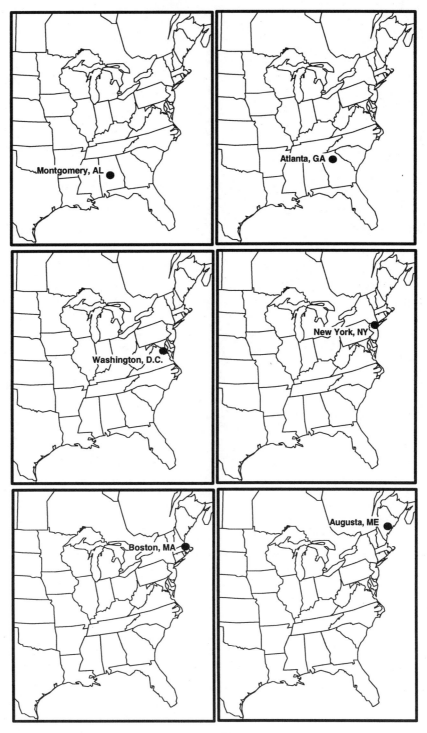

Exercise 10.4 Wind Direction and Nor'easters

The map on the bottom of the page shows the track of a strong cyclone that brought snowfall and hazardous winter weather conditions to the Southern U.S., the East Coast and the Appalachian Mountains. The position of the cyclone is shown for four times. What was the wind direction (N, NE, E, SE, S, SW, W, NW) at locations A, B and C at the times listed on the table below?

LOCATION	TIME	WIND DIRECTION
A:	14 Feb 0000 UTC	_____
A:	14 Feb 1200 UTC	_____
A:	15 Feb 0000 UTC	_____
B:	14 Feb 0000 UTC	_____
B:	14 Feb 1200 UTC	_____
B:	15 Feb 0000 UTC	_____
B:	15 Feb 1200 UTC	_____
C:	14 Feb 1200 UTC	_____
C:	15 Feb 0000 UTC	_____
C:	15 Feb 1200 UTC	_____

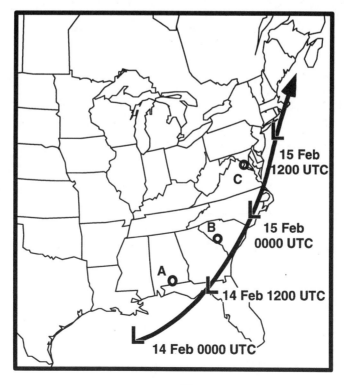

Exercise 11.1 Where Will Freezing Precipitation Occur?

The maps below show analyses of sea level pressure and fronts for four weather systems over the central United States. The shaded areas denote cloud cover. The 0°C isotherm is also indicated on each map. Assume for this exercise that precipitation is falling everywhere that clouds are present. On each of the maps, outline the regions that are likely to be experiencing freezing rain or freezing drizzle.

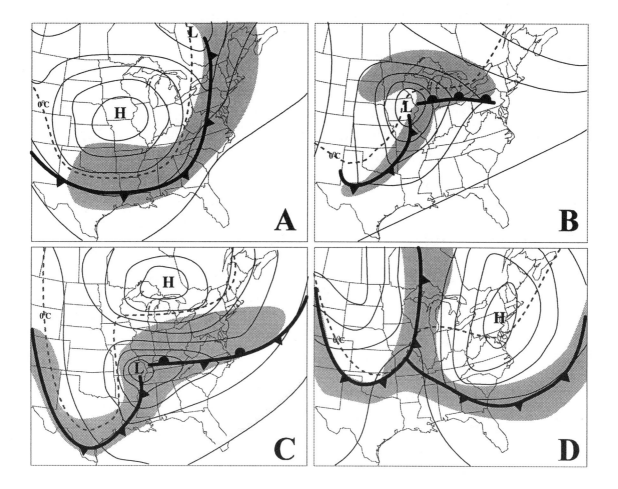

Exercise 11.2 Fronts, Soundings and Precipitation Type

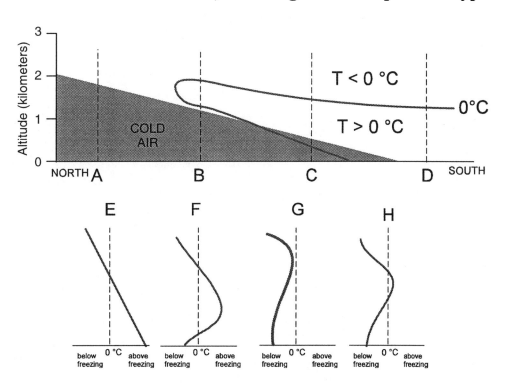

The figure above shows a cross section through a front. Soundings were launched at points A, B, C, and D. The soundings, in no particular order, are shown on diagrams E, F, G, and H. Answer the following questions about the cross section and the soundings.

1. On the cross section, freezing rain formed via the melting process is most likely to occur at point _____.

2. On the cross section, snow is most likely to occur at point _____.

3. On the cross section, rain is most likely to occur at point _____.

4. On the cross section, sleet is most likely to occur at point _____.

5. Which sounding would have been launched in freezing rain that formed via the melting process? _____

6. Which sounding would have been launched in rain? _____

7. Which sounding would have been launched in snow? _____

8. Which sounding would have been launched in sleet? _____

Exercise 11.3 How Much Ice?

Assume that a medium-size tree in your backyard weighs about 1 ton. An early fall cyclone passes your area and produces a local ice storm. Your tree has not yet lost its leaves, and has a full canopy. The freezing rain impacting the tree fell at a light rate, so that about 25% of the rain froze directly on the tree, while the remainder dripped to the ground and ran off, since the temperature of the ground was just above freezing.

Assume that your tree's canopy is circular and has a radius of 5 meters. Also assume that 1 centimeter (0.01 meter, or 0.4 inches) of rain fell during the storm. What percentage of the tree's original weight did the ice on the tree weigh at the end of the storm?

Useful numbers: Density of water = 1000 kilograms/cubic meter

1 ton = 2000 lbs = 908 kilograms.

Exercise 11.4 Freezing Drizzle versus Freezing Rain

Answer each of the following questions below by filling in the blank with either ZR (for freezing rain), ZL (for freezing drizzle), or B (for both freezing rain and freezing drizzle).

1. Can occur in the atmosphere when the temperature of the air aloft is below freezing from the ground to the tropopause. _____

2. Typically is responsible for significant ice storms that produce millions of dollars in damage. _____

3. Can lead to traffic accidents. _____

4. Is the primary cause of aircraft icing. _____

5. Is most common in the Northeastern United States and Canada. _____

6. Is most common on the Great Plains of the U.S. and Canada. _____

7. Occurs when snow aloft falls into a warm layer where the air is above freezing, melts, and then supercools is a subfreezing layer of air near the surface. _____

8. Is composed of supercooled water. _____

9. Does not require a temperature inversion aloft. _____

10. Can freeze upon contact with cars, trees and road surfaces. _____

Name: _____

Section: _____ Date: _____

Exercise 12.1 Lake-Effect Processes

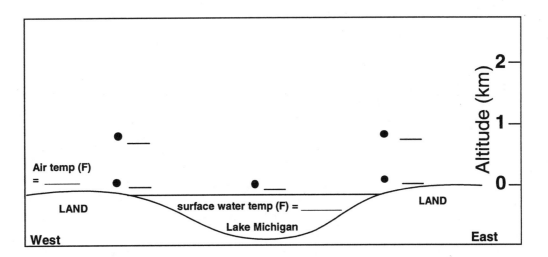

The diagram above shows a cross section across Lake Michigan. It is wintertime, the winds are blowing across the lake from west to east, and a lake effect snowstorm is in progress.

1. Write a reasonable value for the temperature of the lake's surface water on the blank provided on the diagram.

2. Write a reasonable value for the temperature of the air upstream of the lake on the blank provided on the diagram.

3. The statements below each apply to a location indicated by one of the black dots on the diagram. Place the appropriate letter in a blank next to one of the dots.

    A. Convergence occurs at this location.

    B. Divergence occurs at this location.

    C. Air is descending and skies are generally clear at this location.

    D. Air is ascending and heavy snow is falling at this location.

    E. Heat and moisture are transferred from the lake to the air at this location.

4. Draw clouds on this diagram as they might appear during a lake effect storm.

5. Draw arrows to indicate the direction of air motion during a lake effect storm.

Exercise 12.2 Wind Parallel Rolls or a Shore Parallel Band?

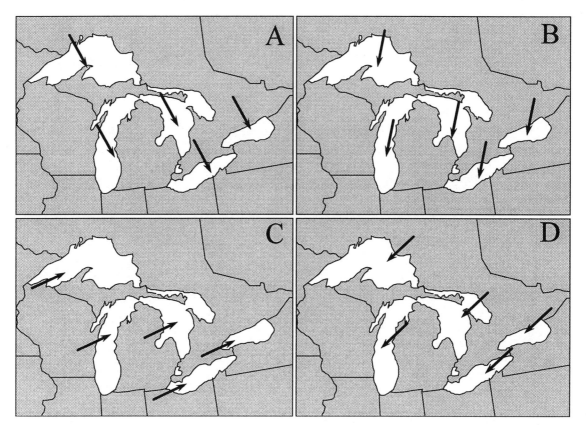

The four maps above each show the Great Lakes region. Assume very cold air is moving across the lakes in the direction of the arrows. Consider each lake on each panel. Would you expect wind-parallel rolls (WPR) or a shore-parallel band (SPB) to form? Answer by filling in the blanks below with the letters WPR or SPB.

PANEL A
Lake Superior _____
Lake Michigan _____
Lake Huron _____
Lake Erie _____
Lake Ontario _____

PANEL B
Lake Superior _____
Lake Michigan _____
Lake Huron _____
Lake Erie _____
Lake Ontario _____

PANEL C
Lake Superior _____
Lake Michigan _____
Lake Huron _____
Lake Erie _____
Lake Ontario _____

PANEL D
Lake Superior _____
Lake Michigan _____
Lake Huron _____
Lake Erie _____
Lake Ontario _____

Name: _____

Section: _____ Date: _____

Exercise 12.3 Lake-Enhanced Snowstorms

Chicago lies on the southwest shoreline of Lake Michigan. Meteorologists often refer to heavy snowfall events in Chicago as "lake-enhanced" rather than "lake-effect" snow. Answer each of the questions below in order to learn what the term "lake-enhanced" means and why it is often more appropriate than "lake-effect" for Chicago snowstorms.

1. Draw an arrow over Lake Michigan that might cause snowfall due to lake-effect processes in the Chicago area.

2. For the wind to blow in the direction you indicated, a low-pressure system is often present. Draw an "L" to indicate the location of a low-pressure system that would lead to the wind you drew. (Remember that air flows counterclockwise around a low-pressure center, spiraling into the low due to friction.)

3. Draw a cold front extending south from the low and a warm front east from the low.

4. Draw a comma-cloud pattern that is consistent with the position of your fronts and the low pressure center.

5. Where is Chicago relative to the comma cloud pattern you drew? _____

6. If the lake was not there, what type of weather conditions would be occurring in Chicago based on its location relative to the cyclone's clouds? _____

7. What does the term "lake-enhanced" mean based on this exercise?

Name: _____

Section: _____   Date: _____

Exercise 12.4 Lake Effect Snow

1.   What direction would the wind need to be in order to get lake effect snow in:

      A. Chicago, IL   _____

      B. Marquette, MI   _____

      C. Gary, IN   _____

      D. Buffalo, NY   _____

      E. Cleveland, OH   _____

      F. Flint, MI   _____

2.   The prevailing wind direction across the Great Lakes is northwest. Shade in where you would expect to find lake effect snow. (Take in to consideration how far inland lake effect snow typically falls.)

3.   Why do you think there is rarely lake effect snow on the northern shores of the lakes?

Exercise 13.1 Surface and Upper Level Maps during Cold Waves

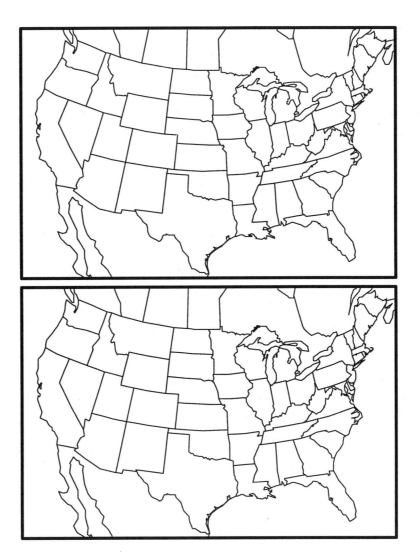

A cold wave is underway. The coldest air is over Missouri. The cold airmass is about one kilometer deep over Missouri.

1. On the top map, sketch the typical 500 mb height pattern (contours) during a cold wave.

2. On the bottom map, sketch the sea level pressure pattern (isobars, every 4 mb) during the coldest days of a cold wave. Also draw a reasonable position for the arctic front.

3. What is a typical value of sea level pressure at the center of the cold airmass during an extreme cold wave?

Name: _____

Section: _____ Date: _____

Exercise 13.2 Wind Chill Temperatures

Wind Chill Chart

The chart above is used by the National Weather Service to determine wind chill temperatures. The table below shows observed values of temperature and wind at ten cities. Fill in the table and determine which city has the coldest wind chill. (Estimate the wind chill temperature by interpolation when necessary.)

City	Temperature (°F)	Wind speed (mph)	Wind Chill Temp (°F)
Bismarck, ND	5	10	
Denver, CO	15	15	
Des Moines, IA	10	10	
Montreal, QE	22	25	
Pittsburgh, PA	19	20	
Boise, ID	5	10	
Chicago, IL	6	15	
Columbus, OH	17	25	
Green Bay, WI	-1	5	
Grand Rapids, MI	1	5	

City with lowest wind chill temperature: _____

Exercise 13.3 Factors Conducive to Cold Waves

Many atmospheric factors, both at the surface and in the upper troposphere, contribute to an extreme cold wave on the Central Plains of United States. Many of these atmospheric factors are listed below. On the map below, place the 2 or 3 letter abbreviation at the geographical location where each of these factors is typically important.

WA: Warm air moves into this area several days prior to a cold air outbreak over the Central Plains.

UC: Convergence occurs in the upper troposphere above this region.

NW: Strong northerly or northwesterly winds are found near the surface during the day or two before the Central Plains reaches its lowest temperatures.

SH: Surface high-pressure center develops and intensifies here.

TM: A "Trigger Mechanism" in the form of a strong extratropical cyclone develops here and tracks northeastward.

SC: Snow cover is more extensive and deeper than usual in this area.

UR: Upper-air ridge has its axis (centerline) in this general area.

CP1: Extremely low temperatures are found in the center of a continental polar high-pressure system in this area several days *prior to* the coldest period on the Central Plains.

CP2: The continental polar high-pressure system might be found in this area several days *after* the coldest temperatures occur on the Central Plains.

Name: _____

Section: _____ Date: _____

Exercise 14.1 Packing the Car for a Blizzard

List as many essential items as you can that you would pack in your car to insure survival if you become stuck on the side of the road in a blizzard.

1. _____

2. _____

3. _____

4. _____

5. _____

6. _____

7. _____

8. _____

9. _____

10. _____

Name: _____

Section: _____ Date: _____

Exercise 14.2 Blizzard Facts

Circle the correct answer in each statement:

1. If caught on the side of a road in your car in a white-out blizzard, you should (place flares at distances of 25, 50, and 100 yards from your car; stay in your car).

2. Ground blizzards are caused by (blowing and drifting; falling) snow.

3. The "Winter of Blizzards" in North Dakota was followed by (a disastrous flood; a large tornado outbreak).

4. In Rocky Mountain cyclones, blizzards occur under the "wrap around" part of the comma cloud which meteorologists refer to as the (trowel; frontal fold).

5. Blizzards associated with Alberta Clippers typically produce (less; more) snow than blizzards associated with cyclones that form east of the Colorado Rockies.

6. Blizzards associated with Alberta Clippers typically have (colder; warmer) temperatures than blizzards associated with cyclones that form east of the Colorado Rockies.

7. In a blizzard, it is likely that exposed skin will feel (warmer; colder) than the wind chill temperature due to melting and evaporation of snow and water on the skin.

8. The National Weather Service (does; does not) use specific temperature criteria to define a blizzard.

9. A "Blizzard Warning" is issued when winds are expected to exceed (15 knots; 30 knots) and falling or blowing snow is expected to reduce visibility to less than a quarter mile for 3 hours.

10. Blizzards in the lower 48 United States occur most often in (North Dakota; Michigan).

Name: _____

Section: _____ Date: _____

Exercise 14.3 Which Type of Blizzard?

Descriptions of blizzard conditions experienced by people who were driving west through North Dakota on Interstate 94 are given below. In each case, identify whether the blizzard described was a Ground Blizzard, a blizzard associated with an Alberta Clipper, or a blizzard associated with a Colorado cyclone. State your reasoning.

1. It was awful. The ground already had 10 inches of snow on it when clouds moved in and it started snowing. The wind picked up and within a hour or two was howling at 40 mph from the north. The snowfall wasn't heavy, but it was blinding. The temperature dropped quickly from 0°F to -20°F. It felt like -50°F! It took us 10 hours to travel 30 miles to the next town. By the time we got there, the weather forecaster on the radio said that about 4 inches of additional snow had accumulated. There was really no way to tell because all of the new snow was fluffy and it drifted in huge piles in some locations, with hardly any in other places.

2. We were driving along on I-94 through a town last night. I could see the moon overhead. Suddenly the visibility dropped to zero. Snow was blowing everywhere. I hit the breaks just in time to keep from crashing into a truck that was pulling off the road. I went on at about 15 mph for a few miles. Just as fast as the snow started it stopped. About 10 minutes later, it happened again. This went on half way across the state, and added several hours to my driving time.

3. Rain was falling when we crossed the Minnesota/North Dakota border heading west. Within about 20 miles the temperature dropped from about 36°F to about 10°F. At first the snow was light, but within about 20 miles, it started really coming down. The snow became so heavy that I couldn't see more than a few feet in front of my car at times. That wasn't the worst of it. The winds picked up when the temperature dropped, and had to have hit 50 mph soon afterward. It was snowing horizontally. We managed to get about 20 miles further to the next exit, only to find that the State Police had closed the interstate. We had no choice but to stay in a nearby motel and wait it out. In the morning when we came outside, everything was covered with at least a foot of snow and drifts were over 5 feet high. It took the highway maintenance crews three days to open the Interstate.

Exercise 14.4 Where Is the Blizzard?

The map below shows a cyclone over the central United States.   Assume it is midwinter and a severe blizzard is occurring.

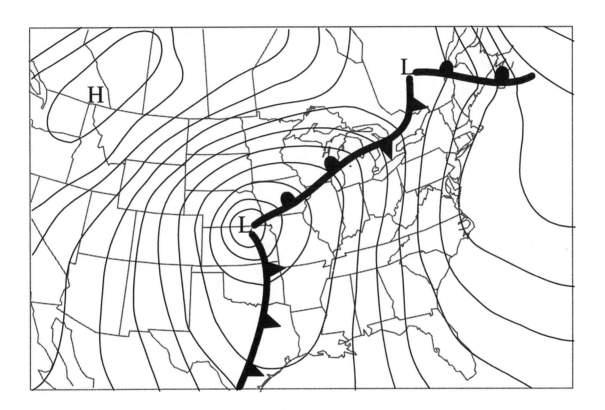

1. Shade in the region of the United States and/or Canada that would be most likely to be experiencing blizzard conditions at the time of the map.

2. What information did you use to determine where the blizzard would be?

3. Is this system an Alberta Clipper or a Colorado cyclone?

Name: _____

Section: _____ Date: _____

Exercise 15.1 Mountain Geography

Sketch the location of each of the listed mountain ranges on the blank map below. List the states in which the mountain ranges are found.

Mountain Range   States in U.S.

Cascades   _____

Sierra Nevada   _____

Coastal Range   _____

Wasatch Range   _____

Bitterroot Mountains   _____

Rocky Mountains   _____

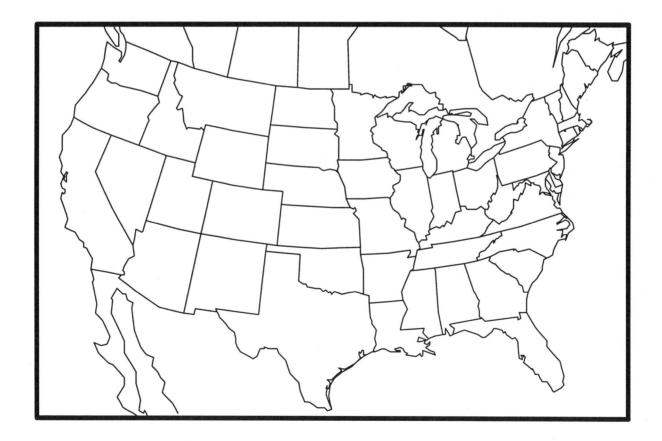

Exercise 15.2 Pressure Patterns, Wind Flow, and Mountain Snows

The four surface maps below each show the sea level pressure distribution across the United States on different days in January. Shade in the regions on each map where the wind direction is favorable for the development of snow in the mountains of the western United States. If the snow would fall on the east slope of the Rockies, label it "upslope snow."

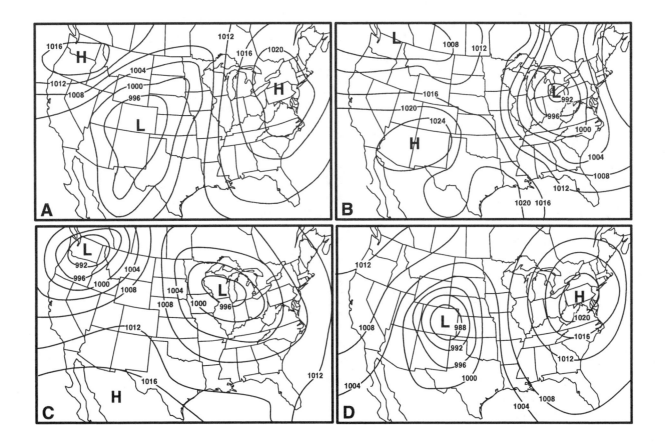

Exercise 15.3 Mountain Snowstorms

Fill in the blanks with either "True" or "False".

1. The water from mountain snowstorms provides about a third of the electricity throughout the western United States and over 80% in Oregon and Washington. _____

2. The fatalities associated with avalanches in the United States have been decreasing in recent years because of better warning systems. _____

3. The strategy used by cloud seeding operations is to convert supercooled water in the clouds to ice crystals. They do this because ice crystals fall faster and have a better chance of reaching the ground than tiny supercooled water droplets. _____

4. The primary reason that mountains experience so many snowstorms in a winter season is that deep cumulonimbus clouds develop over the mountains nearly every day due to heating of the slopes during daytime. _____

5. Many locations along the Sierra Nevada and Cascades receive 20-30 feet (240-360 inches) of snow during a year with average snowfall. _____

6. The Chain Law refers to the chains that are used to block roads during potential avalanches. Drivers are not permitted to take their cars across the chains. _____

7. The water-equivalent of snow in the mountains can range from 4 inches of snow to one inch of water, to 30 inches of snow per inch of water. _____

8. Upslope storms on the east slope of the Colorado Rockies are important because they affect a population corridor that includes the cities of Omaha, Nebraska, and Kansas City, Kansas. _____

9. The weather pattern most conducive to an upslope storm along the east slope of the Colorado Rockies is one in which there is a low pressure system to the north of Colorado and a high pressure system to the south of Colorado.

10. The city of Denver has the best chance of a heavy snowstorm when the surface winds in winter are blowing from the northeast. _____

Name: _____

Section: _____ Date: _____

Exercise 16.1 Characteristics of Mountain Windstorms

Identify the type of mountain windstorm described in each statement by putting the appropriate letter on the blank provided. If the description has more than one correct answer, list all the correct answers on the blank

C – Chinook
S – Santa Ana
K – Katabatic wind

_____ 1. Commonly found on the eastern slopes of the Rocky Mountains.

_____ 2. Found in Greenland and Antarctica.

_____ 3. Can be very strong with wind gusts exceeding 100 knots.

_____ 4. Found in southern California.

_____ 5. Turbulent winds often carry loose snow and create ground blizzards.

_____ 6. Most common during fall and winter.

_____ 7. Can be warm and dry wind similar to an Alpine Foehn or cold and dry,

similar to a Bora.

_____ 8. Caused by cold air that initially develops on top of an elevated landmass.

_____ 9. Can develop when strong a high-pressure system forms over the Great

Basin.

_____ 10. Is capable of quickly melting ice and snow at the base of the mountains.

_____ 11. Winds are always cold in spite of adiabatic warming during descent.

Name: _____

Section: _____ Date: _____

Exercise 16.2 Temperature Changes with Downslope Winds

Determine the air temperature at the base of a mountain, given the height of the mountain and the temperature at mountaintop. Neglect any effects of evaporation or sublimation. Use the space at the right for your calculations. (Be sure to use correct units!)

1. Height of mountain = 3,000 ft (~ 1 km)
   Temperature at mountain top = 0°C
   Temperature at base of mountain = _____

2. Height of mountain = 5,000 ft (~1.5 km)
   Temperature at mountain top = −7°C
   Temperature at base of mountain = _____

3. Height of mountain = 5,000 ft (~1.5 km)
   Temperature at mountain top = −22°C
   Temperature at base of mountain = _____

4. Height of mountain = 8,000 ft (~2.5 km)
   Temperature at mountain top = 5°C
   Temperature at base of mountain = _____

5. Height of mountain = 8,000 ft (~2.5 km)
   Temperature at mountain top = −13°C
   Temperature at base of mountain = _____

6. Height of mountain = 10,000 ft (~3 km)
   Temperature at mountain top = −6°C
   Temperature at base of mountain = _____

7. Height of mountain = 10,000 ft (~3 km)
   Temperature at mountain top = −18°C
   Temperature at base of mountain = _____

Exercise 16.3 Foehn or Bora?

When the Foehn blows down the north side of the Alps in Switzerland, the temperature abruptly warms as cold air is flushed out of the Alpine valleys and replaced with the warm air that descended the mountains. When the Bora blows down the south side of the Dinaric Alps of Yugoslavia, the temperature on the Mediterranean side abruptly drops.

In the United States, temperature changes like the Foehn and Bora can be experienced during downslope windstorms east of the Rockies. The figures below show 3 different temperature conditions prior to the onset a downslope windstorm. Assume in each case that the airmass at the base of the mountain (light shading) will be forced eastward by air descending the mountain from mountaintop. Determine the temperature change of air that descends from the mountain top to the base, and circle whether the temperature change is similar to a Foehn or Bora.

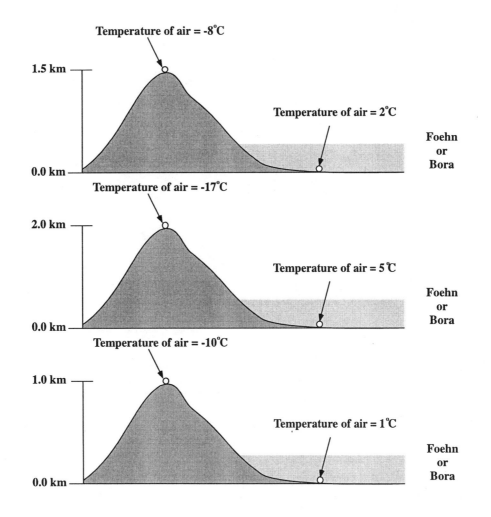

Name: _____

Section: _____   Date: _____

Exercise 16.4 Features of a Downslope Windstorm

On the diagram above place letters where the features listed below are located:

        A.   Chinook Wall

        B.   Rotor

        C.   Hydraulic Jump

        D.   Breaking waves

        E.   Inversion

        F.   Most severe winds at surface

        G.   Shooting flow

        H.   Snowstorm

Exercise 17.1 Characteristics of Thunderstorms

Identify the type of thunderstorm described in each statement.

AM − airmass thunderstorm
SL − squall line
MC − multicell thunderstorm
SP − supercell

_____ 1. Form far from frontal boundaries.

_____ 2. Commonly form along the "tail of the comma cloud.

_____ 3. Appear as a hook shaped echo on radar reflectivity.

_____ 4. Always rotate.

_____ 5. Rarely produce severe conditions.

_____ 6. Occur in clusters.

_____ 7. Can produce hail and tornadoes; most often associated with strong straight line winds.

_____ 8. A low-level jet helps transport warm moist air into the storm and provides low-level wind shear that contributes to storm rotation.

_____ 9. Appear as a long continuous line on radar reflectivity.

_____ 10. Responsible for much of the summer rainfall in the Central Plains.

_____ 11. The anvil can grow to cover an area the size of a U.S state.

_____ 12. Form in regions with weak or no wind shear.

_____ 13. Account for most tornadoes and large hail.

Name: _____

Section: _____   Date: _____

Exercise 17.2 Thunderstorms on Radar and Satellite

On each of the four maps below, for the type of thunderstorm listed:

1. Sketch the approximate size of the cloud shield that would be observed by a satellite.

2. Within the cloud shield you drew, sketch the approximate size of the precipitation region, as would be observed with a radar.

Airmass thunderstorm

Multicell thunderstorm

Squall line

Supercell

Name: _____

Section: _____ Date: _____

Exercise 17.3 Squall Line Structure

1. Identify each of the listed features of a squall line thunderstorm by placing the corresponding letter on the appropriate place on the diagram.

A – anvil

B – bright band on radar (melting level)

C – dry air

D – base of updraft

E – convective region

F – location of gust front at surface

G – heavy rain

H – light rain

I – mammatus

J – moist air

K – overshooting top

L – shelf cloud

M – stratiform region

N – tropopause

O – rear inflow jet

2. Draw five wind barbs at different levels on the vertical line on each side of the storm to show typical wind profiles east of the thunderstorm and west of the thunderstorm. Be sure to illustrate both wind speed and wind direction clearly.

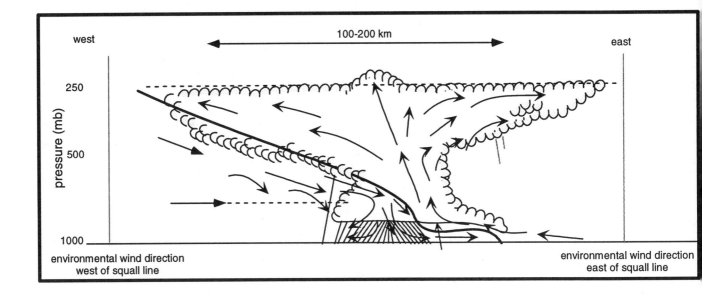

Exercise 17.4 Supercell Thunderstorm Structure

1. Identify each of the listed features of a supercell thunderstorm by placing the corresponding letter at the appropriate location on the diagram.

A – anvil   H – overshooting top

B – bounded weak echo region   I – rear flanking line

C – heavy rain   J – small hail

D –large hail   K – tornado

E – light rain   L – tropopause

F – mammatus   M – virga

G – moderate rain   N – wall cloud

2. Draw several arrows to indicate the location of the updraft and the forward flank downdraft.

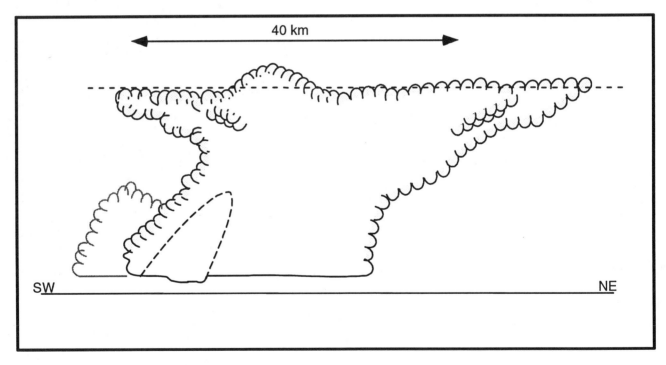

Name: _____

Section: _____   Date: _____

Exercise 18.1 Tornado Wind Speeds

Assume that a tornado is on the ground but has no forward speed. Use the principle of conservation of angular momentum to estimate the wind speeds of the tornado based on the measured rotation of the winds in the mesocyclone. What is the tornado's Fujita Scale rating (use 1 m/s = 2.24 mph)? Use the space to the right of each problem to work through the math.

Conservation of angular momentum states:

$$r_{(mesocyclone)} * v_{(mesocyclone)} = r_{(tornado)} * v_{(tornado)}$$

1. Radius of mesocyclone = 4,000 m
    Rotational velocity of mesocyclone = 2.5 m/s
    Radius of tornado = 100 m
    Rotational velocity of tornado = _____

    Fujita Scale rating:

2. radius of mesocyclone = 1,500 m
    rotational velocity of mesocyclone = 1 m/s
    radius of tornado = 50 m
    rotational velocity of tornado = _____

    Fujita Scale rating:

3. *diameter* of mesocyclone = 3,000 m
    rotational velocity of mesocyclone = 2 m/s
    radius of tornado = 50 m
    rotational velocity of tornado = _____

    Fujita Scale rating:

Name: _____

Section: _____ Date: _____

Exercise 18.2 Tornado-Like Vortices

Identify the vortex described in each of the statements below by choosing from the list provided.

LS – landspout
WS – waterspout
CAF – cold air funnel
GN – gustnado
DD – dust devil

_____ 1. Commonly observed off coastlines in tropical regions.

_____ 2. Forms along the boundary of cool outflow ahead of a thunderstorm.

_____ 3. Develops along fronts that have strong horizontal wind shear.

_____ 4. Are weak, short-lived and are associated with dry convection.

_____ 5. Develop at the base of cumulus cloud associated with a cutoff low.

_____ 6. Sometimes several of these vortices form every few kilometers along a front.

_____ 7. Common over the desert region of the southwestern United States.

_____ 8. Emerge from convective clouds that develop over cool surface air.

_____ 9. Most tornadoes in California are this type.

_____ 10. Rarely reach the ground.

Exercise 18.3 Tornado – Myth or Fact?

Tornadoes are complex and fascinating phenomena. There is much that is not understood about tornadoes and a lot that is *misunderstood* about them. Indicate whether each of the following statements about tornadoes is *fact* or *myth*.

1. Tornadoes have occurred in all 50 states.   MYTH   FACT

2. Tornadoes are attracted to mobile home communities.   MYTH   FACT

3. Tornadoes sometimes sound like a freight train.   MYTH   FACT

4. Tornadoes cannot hit downtown areas because the   MYTH   FACT
    buildings deflect the airflow.

5. Tornadoes typically occur in the part of the thunderstorm   MYTH   FACT
    that is not raining.

6. It is common for the sky to look greenish before a tornado. MYTH   FACT

7. Tornadoes always rotate counter-clockwise in the   MYTH   FACT
    Northern Hemisphere.

8. If a tornado watch has been issued, then a tornado has   MYTH   FACT
    been sighted.

9. If there is a tornado warning, you should open all the   MYTH   FACT
    windows in your house or apartment.

10. When the tornado siren goes off, you should grab your   MYTH   FACT
    video camera and go outside to catch it on film.

11. Tornadoes can be made up of 2 or more smaller tornadoes   MYTH   FACT
    called suction vortices.

12. If you are driving in a car on the highway and see a   MYTH   FACT
    tornado behind you approaching rapidly then you should
    try to outrun it.

13. If you are driving in a car on the highway and see a   MYTH   FACT
    tornado behind you, find the nearest overpass and hide
    under it.

14. When the tornado siren goes off, go to the innermost   MYTH   FACT
    room on the lowest level of your house or building.

Name: _____

Section: _____   Date: _____

Exercise 18.4 Probability of Experiencing a Tornado

Suppose that the National Weather Service issues a tornado watch for an area containing your location. Assume a typical size of a tornado "watch box" is 100 miles × 200 miles.

1. Suppose that a single tornado actually occurs within this watch box area. What is your probability of being struck by the tornado in each of the following scenarios of tornado occurrence?

	F rating	*length of path*	*width of path*	*probability of being struck*
1.	F1	2 mi.	0.1 mi.	_____
2.	F2	5 mi.	0.2 mi.	_____
3.	F3	10 mi.	0.25 mi.	_____
4.	F4	20 mi.	0.3 mi.	_____
5.	F5	40 mi.	0.4 mi.	_____

2. What is the probability that you would be within 25 miles of a tornado if one tornado occurred in the watch box?

3. Suppose that the F5 tornado in (1) only had F5 winds for 1 mile of its path length and only on the right half of the damage path. What is the probability that you would be struck by the tornado with full F5 intensity?

70

Name: _____

Section: _____ Date: _____

Exercise 19.1 Hailstone Sizes

Idealized cross-sections of various hailstones, drawn to scale, are shown below. (a) Each hailstone is a different size. Identify the hailstone as: pea, marble, quarter, golf ball or tennis ball sized; (b) Record the approximate diameter of each in centimeters and inches; (c) estimate the number of supercooled droplets that are contained in each hailstone. Assume the diameter of a supercooled droplet is 0.2 millimeters. The volume of a sphere is $V = 4/3\ \pi\ r^3$ or $1/6\ \pi\ d^3$, where $\pi = 3.14$, r is radius, and d is diameter. (Your answers for the top hailstone should go on the first line of blanks, the next on the second line, and so on.)

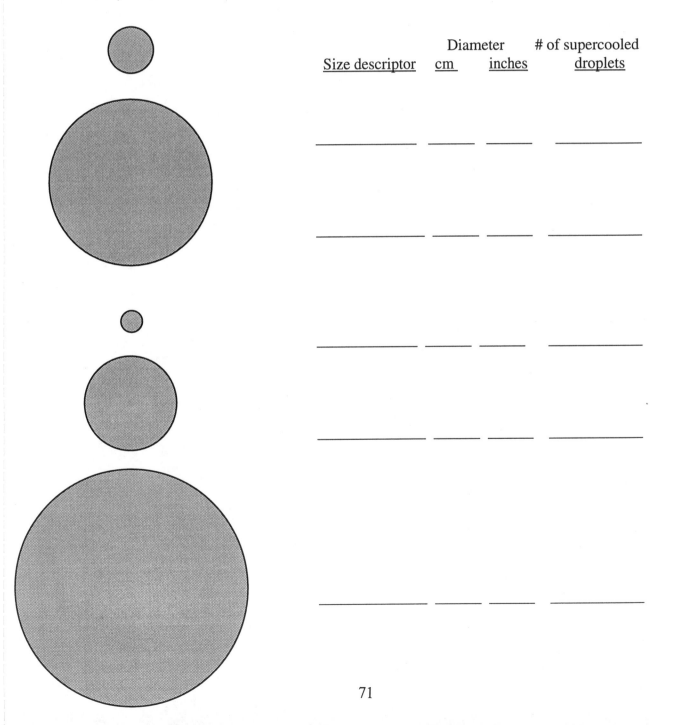

	Size descriptor	Diameter cm	inches	# of supercooled droplets
	_____	____	____	_____
	_____	____	____	_____
	_____	____	____	_____
	_____	____	____	_____
	_____	____	____	_____

Name: _____

Section: _____ Date: _____

Exercise 19.2 Hail Distribution in Thunderstorms

The diagram below is a plan view of radar reflectivity from a supercell thunderstorm. Suppose that three cars drive through the thunderstorm along transects AA', BB' and CC', starting in each case at the unprimed letter.

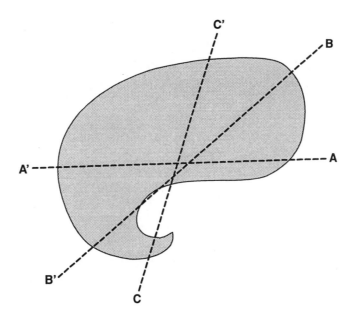

1. List the precipitation types the driver would encounter on each route, in the order they would appear along the route.

   A'_____A

   B'_____B

   C'_____C

2. Place an "X" on each transect in the diagram at the location where a car is in the greatest danger of experiencing hail damage.

Name: _____

Section: _____   Date: _____

Exercise 19.3 Polarization Diversity Radars and Hail Detection

Polarization diversity radars measure both reflectivity and differential reflectivity. The magnitude of the reflectivity is related to the size and number of raindrops and hailstones intercepted by the radar beam, while the differential reflectivity is related to the shape of the raindrops and hailstones.

Very small raindrops and cloud droplets are spherical. On average, most hailstones are also spherical. For these spherical objects, the differential reflectivity is between 0-1 decibels. Large raindrops flatten as they fall into shapes resembling hamburger buns. For large raindrops, the differential reflectivity is about 1-3 decibels.

The reflectivity has low values (10-20 dBZ) in parts of the storm containing very small water droplets, moderate values (20-40 dBZ) in parts of the storm containing rain, and high values (40-60 dBZ) in parts of the storm that may either have heavy rain or hail.

Used together, the reflectivity and the differential reflectivity can be used to isolate parts of storms containing only hail. The cross sections below show the reflectivity and differential reflectivity measured in the bottom half of a supercell thunderstorm. Using the criteria above, draw a line (on both diagrams) enclosing the region that is likely to contain hail. In the space below the diagram, state your reasoning.

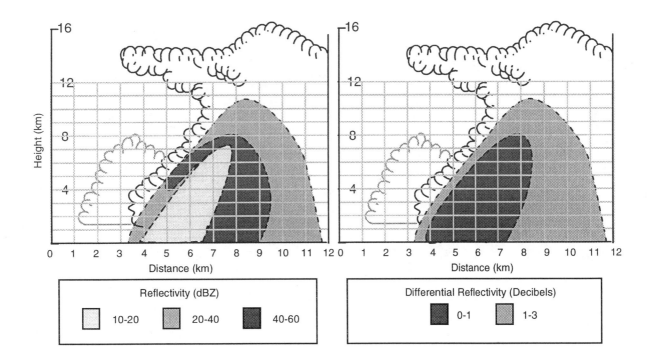

Name: _____

Section: _____ Date: _____

Exercise 19.4 Hailstreaks and Hailswaths

The white regions on the diagrams below show hailstreaks that occurred during two severe thunderstorm outbreaks. In each case, outline the hailswaths and complete the table below.

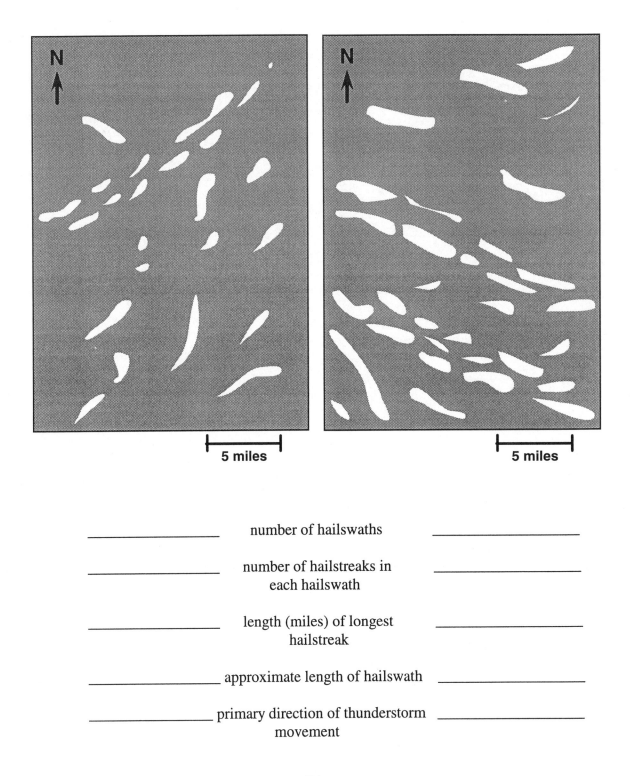

5 miles   5 miles

_____ number of hailswaths _____

_____ number of hailstreaks in
                   each hailswath _____

_____ length (miles) of longest
                   hailstreak _____

_____ approximate length of hailswath _____

_____ primary direction of thunderstorm
                   movement _____

Exercise 20.1 Lightning – Myth or Fact?

Determine whether each of the statements below is myth or fact with regard to lightning and lightning safety.

1. Lightning is sometimes associated with thunder, but not always.   MYTH   FACT

2. Lightning never strikes the same place twice.   MYTH   FACT

3. Your personal risk of being killed by lightning is about 1 in a million.   MYTH   FACT

4. If you are caught outdoors during an electrical storm, one of the safest places to be is in your car.   MYTH   FACT

5. If it is not raining, then there is no danger from lightning.   MYTH   FACT

6. The rubber soles of shoes will protect you from being struck by lightning.   MYTH   FACT

7. People struck by lightning carry an electrical charge and should not be touched.   MYTH   FACT

8. "Heat lightning" occurs after very hot summer days and poses no threat to anyone when it occurs.   MYTH   FACT

9. A lightning stroke heats the air to a temperature five times the temperature of the sun's surface.   MYTH   FACT

10. Lightning can be predicted.   MYTH   FACT

11. The average flash of lightning will light a 100 watt bulb for more than 3 months.   MYTH   FACT

12. It is unsafe to be indoors near appliances or plumbing during a lightning event.   MYTH   FACT

13. If you feel your hair stand on end during a storm, crouch down low to the ground – you are about to be struck by lightning!   MYTH   FACT

Exercise 20.2 Lightning Phenomena

Match the statements to the types of lightning phenomena listed below. Some answers can be used more than once.

> HL – heat lightning
> BE – bead lightning
> SL – sheet lightning
> SE – St. Elmo's Fire
> RS – red sprites
> BJ – blue jets
> EL – elves
> BA – ball lightning
> PP – positive polarity lightning
> ST – stepped leader

_____ 1. Earliest part of a lightning discharge in which negative charge begins to descend below cloud base.

_____ 2. Flash of light overhead from a lighting stroke so far in the distance that thunder is not heard.

_____ 3. Large, weak luminous flashes that emanate upward from the anvil regions of thunderstorms and are brightest at heights of 65 – 75 km.

_____ 4. Cannot be detected by the naked eye, but extend in a cone shape from the top of the active part of thunderstorms.

_____ 5. Sparks that occur on metal objects.

_____ 6. Lightning from the anvil region of a thunderstorm.

_____ 7. Floats in air after a lightning strike; ranges in color and may last 30 seconds.

_____ 8. Disk-shaped regions of light found far above a thunderstorm.

_____ 9. Lightning stroke breaks into separate, distinct segments of light that do not last long enough to be observed by the human eye.

_____ 10. Illuminates a cloud uniformly.

Exercise 20.3 Lightning in the United States

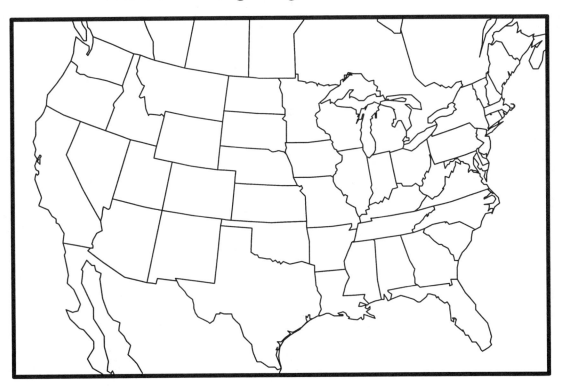

1. You are giving a talk about lightning to a 6th grade class. The teacher of the class brings out the map above and asks you to show the children where in the United States people are in (a) the greatest danger, (b) moderate danger, (c) little danger (d) almost no danger to be stuck by lightning. Sketch on the map above your answer to this question.

2. Give the children a simple explanation for your answer (write your answer below).

Exercise 20.4 Lightning Development

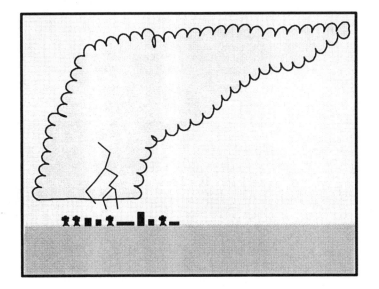

Shown above is a thunderstorm cloud that is in the first stage of development of a lightning discharge.

1. On the diagram, show the distribution of charge within the cloud and at the ground just prior to the lightning stroke. (Use the symbols "+" and "-" for the positive and negative charges, respectively).

2. What is the name given to the descending region of negative charge shown in the figure?

3. Describe briefly the remaining sequence of events that will occur during the lightning discharge.

Name: _____

Section: _____   Date: _____

Exercise 21.1 Aircraft and Downbursts

Shown below is a cross-section through a downburst. The horizontal wind speeds in the plane of the diagram are shown at various altitudes.

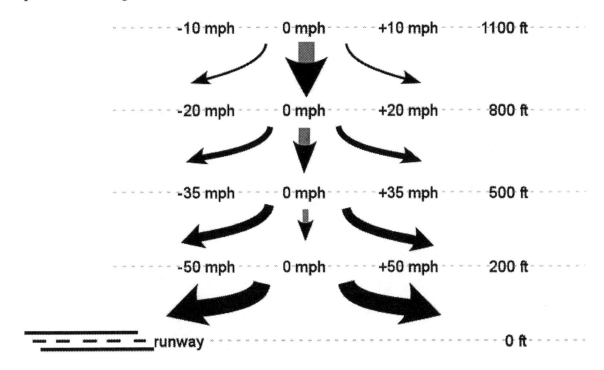

1. A single-engine Cessna plane enters the downburst from the right at an elevation of 500 feet. The plane's air speed in the calm air outside the downburst is 150 mph. How will the airspeed change as the pilot flies through the downburst if the pilot maintains a ground speed of 150 mph at all times and manages to keep the plane at exactly 500 feet above the earth's surface?

2. Assume instead that the pilot does not maintain a constant altitude, but does maintain a constant ground speed of 150 mph. Sketch on the diagram above a plausible scenario for the plane's altitude as it flies through the downburst.

3. Suppose the same plane enters the downburst from the left instead of the right. How does your answer to (2) change?

4. Suppose the same plane is preparing to land from right to left on the runway shown above. The plane's stall speed is 100 mph. If the plane's ground speed is 130 mph in the calm air to the far right of the downburst and the pilot maintains this speed relative to the ground, how high in the downburst does the plane have to be in order to avoid stall speed in the approach to the runway? What should the pilot do?

Name: _____

Section: _____ Date: _____

Exercise 21.2 Downburst Indications from Soundings

Shown below are two soundings.

1. Compare the soundings in terms of four characteristics of an atmospheric environment conducive to downbursts by completing the table at the bottom of this page.
2.
3. Determine which region (A or B) is more likely to experience downbursts.

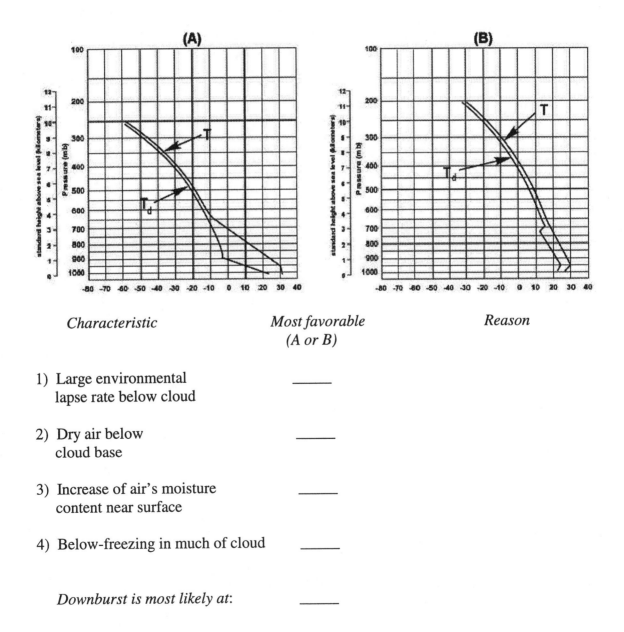

Characteristic	Most favorable (A or B)	Reason
1) Large environmental lapse rate below cloud	_____	
2) Dry air below cloud base	_____	
3) Increase of air's moisture content near surface	_____	
4) Below-freezing in much of cloud	_____	
Downburst is most likely at:	_____	

Name: _____

Section: _____   Date: _____

Exercise 21.3 Downburst Detection with Doppler Radar

The schematic below is a plan view (top down) of the winds associated with a downburst. A stationary downburst's core, denoted by "**C**", is surrounded by outflow as shown. A Doppler radar is located at point "**R**", east of the downburst.

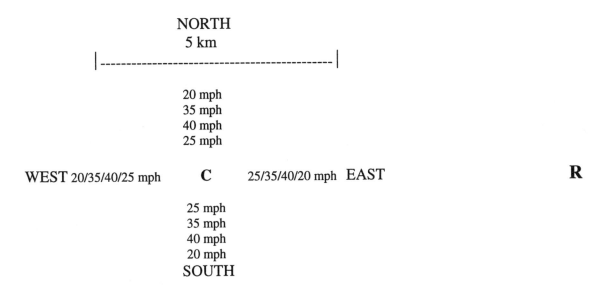

1. In the space below, draw schematically the radial velocity field observed by the radar. Assume that the winds in the environment surrounding the downburst are calm.

2. Suppose the same downburst is traveling *eastward* (rightward) at 20 mph, the speed and direction of the surrounding winds. What will be the maximum winds in each quadrant?

    North: _____ East: _____ South: _____ West: _____

    In the space below, sketch the radial velocity field measured by the radar.

3. Suppose the same downburst is traveling *westward* (leftward) at 20 mph, the speed and direction of the surrounding winds. What will be the maximum winds in each quadrant?

    North: _____ East: _____ South: _____ West: _____

    Sketch the radial velocity field measured by the radar.

Name: _____

Section: _____   Date: _____

Exercise 21.4 Downburst Detection with Surface-based Anemometers

One method used near airports to detect downbursts is a network of closely spaced anemometers that are connected to a computer programmed to detect the spatial patterns of downburst winds. This system is called the Low-Level Windshear Alert System (LLAWS). This exercise illustrates in a simplified manner the basis of downburst detection by LLAWS.

Shown below is a network of surface anemometers, surrounding a north-south airport runway. The anemometers are labeled **A-F**. A circular downburst, with its center at the black dot on the left moves from left to right at a speed of 12 mph across the anemometer network between 1600 and 1630 local time. The downburst's diameter is 2 miles, and its *winds are radially outward at 40 mph at all points within the 2-mile diameter circle*. (This is an idealization, since the wind speeds of an actual downburst will vary across the downburst area). Assume that winds in the background environment are calm.

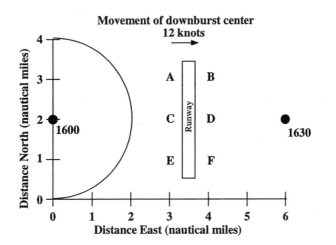

1. Using standard wind symbols (staffs and barbs), show in the table below how the winds varied between 1600 and 1630 local time at the different anemometers

	A	B	C	D	E	F
1600	____	____	____	____	____	____
1615 (downburst center at C)	____	____	____	____	____	____
1620 (downburst center at D)	____	____	____	____	____	____
1630	____	____	____	____	____	____

2. If the LLAWS software is designed to sound an alarm when the wind speed differs by 20 mph between any two anemometers, at what time should an LLAWS downburst alarm have sounded?

Name: _____

Section: _____   Date: _____

Exercise 22.1 Impacts of El Niño and La Niña

The table below is a summary of weather statistics during recent El Niño and La Niña years.

	Year	Winter precipitation in San Francisco, CA	Total snowfall in Urbana, IL	Number of hurricanes in Atlantic Ocean
Average		19.7"	26"	6
El Niño years	1972-1973	27.1"	5"	3
	1982-1983	25.1"	15"	2
	1991-1992	15.2"	10"	4
	1997-1998	37.3"	12"	3
La Niña years	1973-1974	14.5"	32"	5
	1988-1989	12.3"	24"	12
	1995-1996	21.4"	38"	11
	1998-1999	15.5"	29"	10

1. What relationship do you see between winter precipitation in San Francisco and El Niño/La Niña years?

2. What relationship appears to exist between snowfall in Urbana, IL and El Niño/La Niña years?

3. What relationship appears to exist between hurricanes in the Atlantic Ocean and El Niño/La Niña years?

4. The actual precipitation amount (rain and snow together) in Urbana, Illinois during El Niño years is about average. Why would there be less snow?

5. During El Niño years, the polar and subtropical jetstreams are often in a different location compared to other years. Why would this cause fewer hurricanes in the Atlantic Ocean and more precipitation in San Francisco?

Name: _____

Section: _____   Date: _____

Exercise 22.2 ENSO and the Tropical Atmosphere and Ocean

Label the following statements True (T) or False (F).

1. _____ The Walker Cell is a north-south circulation extending from the surface to the upper troposphere.

2. _____ The term "Southern Oscillation" refers to the east-west seesaw of sea level pressure in the tropical Pacific Ocean.

3. _____ Tahiti is normally dominated by low pressure at the surface.

4. _____ During the El Niño phase of the Southern Oscillation, the normal upward vertical motion over Darwin is reduced in intensity.

5. _____ Upwelling normally occurs along the western coast of Peru and Ecuador.

6. _____ Upwelling replaces ocean surface water with warmer water.

7. _____ Upwelling is stronger during a La Niña than during an El Niño.

8. _____ When the Walker Cell is stronger than normal, the trade winds are stronger than normal.

9. _____ Ocean surface temperatures normally increase westward in the tropical Pacific Ocean because the trade winds blow from east to west.

10. _____ When the Walker Cell weakens, the sea level pressure over Indonesia and northern Australia decreases.

11. _____ The eastern tropical Pacific receives more rain during a La Niña than during an El Niño.

12. _____ In the trade wind region, the winds in the upper troposphere generally have an eastward component.

13. _____ Winds near the equator are largely a response to the pressure gradient force because the Coriolis force is weak.

Name: _____

Section: _____   Date: _____

Exercise 22.3 The Southern Oscillation Index (SOI)

The Southern Oscillation is based on the difference between the monthly sea level pressures at Tahiti in the central Pacific and Darwin, Australia. Specifically, the pressure's departure from normal at Darwin (PDN $_{Darwin}$) is subtracted from the pressure's departure from normal at Tahiti (PDN $_{Tahiti}$):

$$Southern\ Oscillation\ Index\ =\ PDN\ _{Tahiti}\ -\ PDN\ _{Darwin}$$

SOI values of -1 and +1 are generally regarded as the thresholds for El Niños and La Niñas, respectively.

The table below contains actual values of the departures from normal pressure and/or values of the Southern Oscillation Index from various months during the past 20 years. In each case, fill in the missing value and indicate whether an El Niño or a La Niña (or neither) is occurring.

	P_{Tahiti}	P_{Darwin}	SOI	El Niño or La Niña?
Jan. 1983	-2.8	+3.0	_____	_____
Jan. 1984	+0.4	_____	+0.1	_____
Mar. 1987	-1.1	+2.0	_____	_____
Jul. 1988	+1.5	-0.2	_____	_____
Jan. 1992	-1.3	+1.7	_____	_____
Jan. 1993	_____	+1.3	-1.3	_____
Jun. 1997	_____	+2.2	-2.8	_____
Nov. 1998	+1.1	_____	+1.4	_____
Jan. 2001	-0.2	+0.4	_____	_____
Aug. 2002	-1.0	+0.6	_____	_____

85

Name: _____

Section: _____   Date: _____

Exercise 22.4 ENSO and the Subtropical Jetstream

El Niño events affect the heating of the tropical and subtropical atmosphere. Variations in this heating affect the upper air pressures and the subtropical jetstream. The subtropical jetstream, in turn, links the tropical heating to the weather of North America, especially the southern United States and the North Atlantic Ocean.

The three diagrams below represent north-south vertical cross sections through the eastern tropical and subtropical Pacific Ocean, in the general vicinity of 110°-130°W, which is the range of longitudes centered on California.. The sea surface temperatures are indicated, respectively, as (A) near-normal, (B) above-normal and (C) below-normal. In each case, provide schematic representations of clouds (size proportional to amount, if present). Also draw a single line to represent the 250 mb surface (sloped in a manner to account for the subtropical jetstreams in each hemisphere. Show the location of the subtropical jetstream by an "SJ", and make the size of the "SJ" proportional to the intensity of the subtropical jetstream.

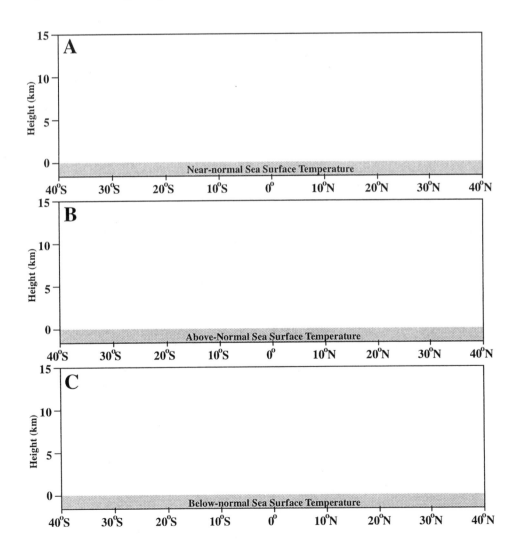

Name: _____

Section: _____ Date: _____

Exercise 23.1 Hurricane Structure

The top diagram below is a cross-section of the clouds and circulations within a strong hurricane. The shaded areas represent the eyewall and spiral rainbands. In the lower portions of the diagram, show schematically how the surface pressure, surface wind speed, rainfall rate, 700 mb temperature and storm surge height vary across the storm. Also, use the graph to the right of the cross-section to show how the wind speed varies with altitude in the eyewall.

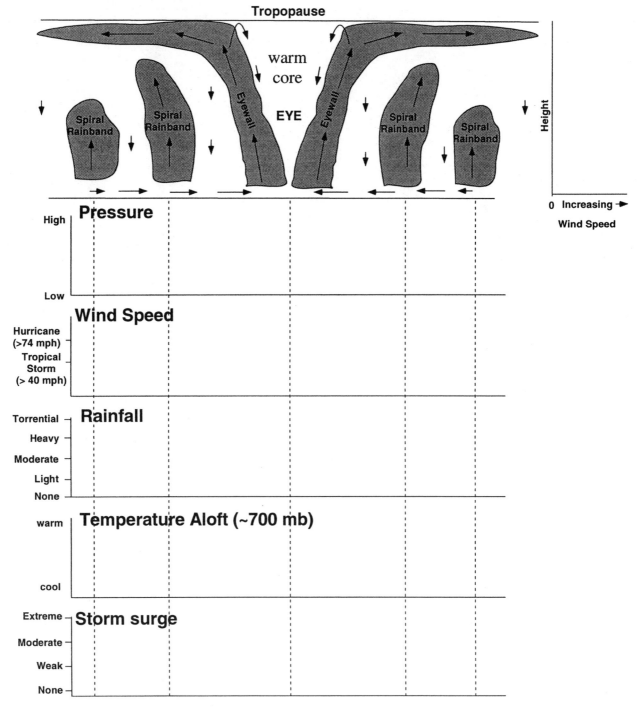

Name: _____

Section: _____   Date: _____

Exercise 23.2 Hurricane Prediction

A hurricane is located in the Gulf of Mexico at 6 pm EDT on September 1. The exact location is shown by the hurricane symbol on the map below. Information about the forecast speed and direction of movement is provided below. The hurricane's diameter is approximately 150 miles.

Time/Date	Rotational Wind	Forward Speed	Moving Toward	Saffir-Simpson Rating
6 pm Sep. 1	125 mph	30 mph	NE	_____
6 am Sep. 2	105 mph	25 mph	NE	_____
6 pm Sep. 2	105 mph	15 mph	N	_____
6 am Sep. 3	80 mph	12 mph	E	_____

1. Complete the table above by determining the Saffir-Simpson rating at each time.

2. Plot the hurricane's track through 6 pm September 3 on the map above.

3. At what date and time will the hurricane make landfall? _____

4. What will be the storm's strongest wind speed at landfall? _____

5. Shade in the region that will experience the highest storm surge as the hurricane makes landfall.

6. Place a small "x" in the region that will experience the highest storm surge as the hurricane moves out over the Atlantic Ocean.

7. Provide a reason why the hurricane strength is predicted to change through Sep. 3.

8. Describe the sequence of winds (speeds and directions that can be expected at Jacksonville, FL as the hurricane passes that location.

9. Why do you think the hurricane tracked in the direction it did?

Name: _____

Section: _____ Date: _____

Exercise 23.3 Hurricane Winds and Saffir-Simpson Ratings

Information about the rotational wind speeds and forward movement of four hurricanes is provided in the sketches below. Four quadrants in each storm are defined relative to the direction the storm is moving. Use the information to evaluate the winds in the eyewall in the four quadrants of the storm (right, left, front and back). In addition, use the information provided to determine the Saffir-Simpson rating of each hurricane.

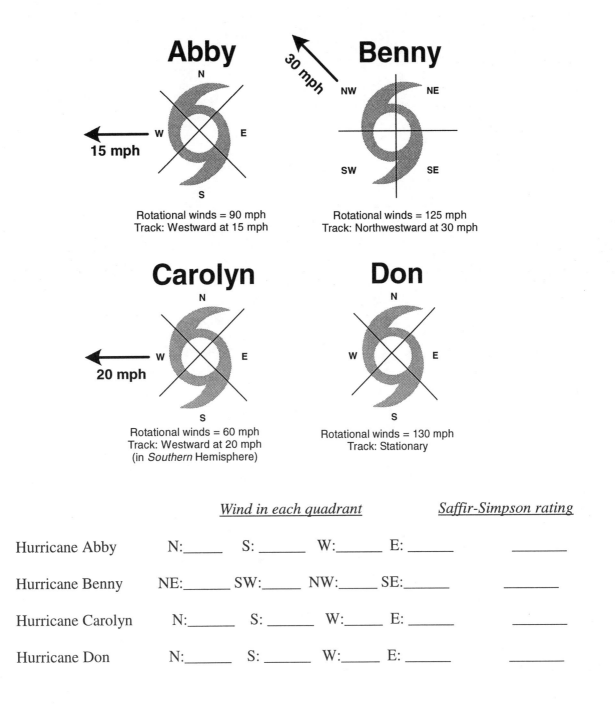

Abby

Rotational winds = 90 mph
Track: Westward at 15 mph

Benny

30 mph

Rotational winds = 125 mph
Track: Northwestward at 30 mph

Carolyn

20 mph

Rotational winds = 60 mph
Track: Westward at 20 mph
(in *Southern* Hemisphere)

Don

Rotational winds = 130 mph
Track: Stationary

	Wind in each quadrant				*Saffir-Simpson rating*
Hurricane Abby	N:_____	S: _____	W:_____	E: _____	_____
Hurricane Benny	NE:_____	SW:_____	NW:_____	SE:_____	_____
Hurricane Carolyn	N:_____	S: _____	W:_____	E: _____	_____
Hurricane Don	N:_____	S: _____	W:_____	E: _____	_____

Exercise 23.4   Tropical Cyclones: Physical and Dynamical Processes

Classify the following as either triggers of tropical thunderstorms (T), requirements of the environment for tropical thunderstorms to organize into hurricanes (E), or mechanisms by which a tropical cyclone's winds intensify (I). Also circle the word in parentheses that best captures the relationship.

_____ Low-level (divergence, convergence) in an easterly wave in the trade winds.

_____ Location (more than, less than) five degrees from the equator.

_____ Ocean surface temperatures (above, below) a threshold value.

_____ Angular momentum (generation, dissipation, conservation).

_____ Rising motion in the (Intertropical, Subtropical, Subpolar) Convergence Zone.

_____ (Strong, Weak) vertical wind shear.

_____ Mid-latitude (cold, warm) fronts that reach the tropics.

_____ Wind-induced transfer of heat from (atmosphere, ocean) to (atmosphere, ocean).

_____ (Shallow, Deep) layer of warm water in the upper ocean.

_____ Rotation of winds such that distance from axis of rotation (increases, decreases).

_____ (Increase, Decrease) of surface pressure in the storm's center.

_____ Winds that (increase rapidly, do not change) with height.

Name: _____

Section: _____ Date: _____

Exercise 24.1 Flood Types

Choose one of the following options that best fits each statement below.

    A.   Coastal flood
    B.   Widespread flood
    C.   Flash flood

1. Can be made worse if it occurs during a full moon. _____

2. The time and day of the worst flooding is often predictable several days in advance.

    _____

3. Is the most deadly. _____

4. Often enhanced by snowmelt. _____

5. Common in the western mountain regions. _____

6. Occurs most often in July. _____

7. Is associated with Storm Surge. _____

8. Occurs on large rivers. _____

9. More frequent on small rivers and streams. _____

10. Most costly. _____

Name: _____

Section: _____ Date: _____

Exercise 24.2 Weather Phenomena and Floods

Seven weather phenomena associated with floods are listed below. Below the list are descriptions of floods taken from news reports. Match the report with the weather phenomenon that caused the flood described in the report. Each flood type is used once.

A. Tropical cyclone after landfall   E. Squall Line
B. Multicell thunderstorm complex   F. Thunderstorms over mountains
C. Pineapple Express   G. Floods enhanced by snowmelt
D. Frontal overrunning

1. July 26: Three campers were killed today as a flash flood roared down a canyon in which they were camping and swept them into the rocks below the campground. Witnesses said that the flood came out of nowhere. Only light rain was reported at the campground at the base of the canyon, although lightning flashes and occasional distant thunder were observed the entire night just to the northeast. ____

2. November 30: Residents along the Little Sandy River in eastern Kentucky were forced from their homes after four days of steady cold rain caused the river to overflow its banks and flood homes along the river. After 15 inches of rain, insult was added to injury as snow fell on the beleaguered residents. _____

3. September 20: Torrential rain fell today in the eastern valleys of the Appalachian mountains, causing very heavy flooding along both minor and major river systems. The rain was accompanied by buffeting winds, which felled trees and kept rescuers from reaching some remote areas that were particularly hard hit by flooding. _____

4. February 12: Over 30 inches of rain fell over 3 days in the mountains as southwesterly winds continued for yet another day. Reservoirs, unable to store the water, were forced to discharge it into the levee system. Several levees were topped and failed, causing three towns to fill with water to near their rooftops. _____

5. July 5: Extremely heavy rain accompanied by thunder and lightning ruined the July 4 celebration as storm drains filled to capacity and backed water into sewage systems. Forecasters initially expected the rain to move on, but the cold front stalled according to one forecaster, causing a "train" of thunderstorms to track over the city. _____

6. March 21: The flood is the result of a long winter, followed by heavy spring rains over western Pennsylvania and West Virginia. The crest on both the Allegheny and Monongahela rivers passed Pittsburgh last week and is now working its way down the Ohio River, where it is expected to pass Cincinnati in two days. _____

7. August 1: Extremely slow moving thunderstorms rumbled across western Iowa overnight. The storms continually regenerated over the same area for twelve hours, causing flooding along the Raccoon, Des Moines and Nishnabotna rivers. _____

Name: _____

Section: _____   Date: _____

Exercise 24.3 Weather Maps Associated with Flood Events

Floods occur in Omaha, Nebraska in mid July (Map A), Mobile, Alabama in mid September (Map B), and Nashville, TN in late December (Map C).

1. Draw plausible surface maps showing the positions of low and high pressure centers, fronts or other important meteorological features that may have triggered each of these three floods.

2. Identify the weather system that caused of the flood based on the map you drew.

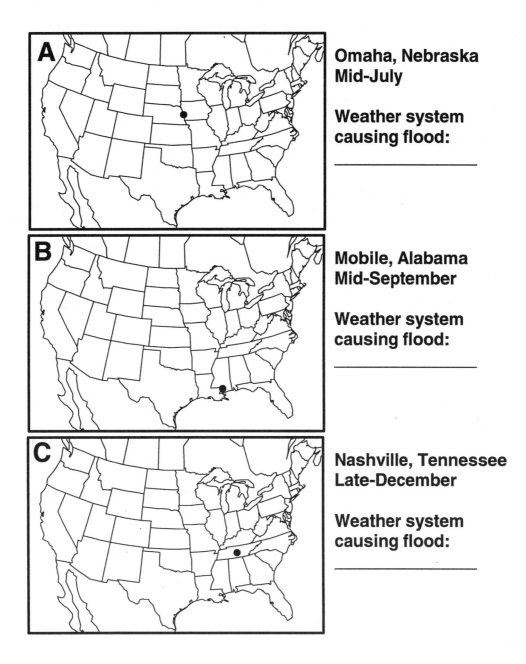

Omaha, Nebraska Mid-July

Weather system causing flood:

_____

Mobile, Alabama Mid-September

Weather system causing flood:

_____

Nashville, Tennessee Late-December

Weather system causing flood:

_____

Name: _____

Section: _____ Date: _____

Exercise 24.4 Flood Safety and Preparedness

Label each statement about flood safety and preparedness as either T (True) or F (False).

1. _____ If your home is inundated by a 100-year flood in 2004, you can safely assume that a similar flood will not occur for the rest of the 21st century.

2. _____ An automobile can be swept away by as little as twelve inches of water.

3. _____ The skill of the National Weather Service in identifying potential flash flood events increased significantly in the 1990s as a result of the deployment of NEXRAD radars.

4. _____ The National Weather Service issues a flash flood watch when a flash flood is occurring or imminent.

5. _____ A levee protects a flood plain by decreasing a river's flow rate.

6. _____ The actual amount of water that causes a 100-year flood varies from river to river and along a particular river.

7. _____ Flash floods had little impact on the 1993 flood of the Mississippi River.

8. _____ Snowmelt is responsible for most flash floods that occur in the United States.

9. _____ Not all factors affecting flood intensity are weather-related.

10. _____ Widespread floods are sometimes described as "leisurely disasters."

11. _____ In the United States, the peak in flash flood activity occurs in July, the time of year when thunderstorms are most common.

12. _____ Because river conditions are monitored using streamflow gauges and water depth monitoring systems, the forecasting of flash floods is no longer difficult.

13. _____ The National Weather Service Flood issues flood watches and warnings on a county-by-county basis.

14. _____ Coastal flooding occurs only in association with tropical cyclones.

15. _____ If your vehicle stalls on a flooded roadway, do not abandon the vehicle.

16. _____ It is more difficult to recognize the signs of flooding at night.

Exercise 25.1 Types and Impacts of Drought

Several types of drought are distinguished on the basis of the impacts of a rainfall deficit. Examples of drought impacts are provided below. In each case, indicate whether the impact is best described as a manifestation of:

M: Meteorological drought (Lack of rainfall)
H: Hydrological drought (Reduced streamflow and reservoir capacity)
A: Agricultural drought (Drought reduces crop yield and impacts farming)
S: Socioeconomic drought (Financial losses to people and industries beyond farming).

1. _____   Streamflow rates drop to their lowest levels in 50 years.

2. _____   Topsoil is picked up by the wind and blown across several states.

3. _____   Rainfall for the spring season is less than 50 percent of normal over the Southern Plains.

4. _____   The price of future corn contracts on the Chicago Board of Trade rises to the maximum allowable amounts for five consecutive days in July.

5. _____   Thousands of residents are driven to migrate westward from the Plains after five years of drought.

6. _____   Barge traffic comes to a halt on the Mississippi River.

7. _____   Ski-area operators in the western United States suffer through their worst year in a decade because of a deficient snow cover.

8. _____   Trees are severely stressed during a hot dry summer, resulting in the death of an abnormal percentage of trees over the next year.

9. _____   A local municipality imposes a ban on the washing of cars and the watering of lawns.

10. _____   Foundations of buildings are damaged as the surrounding soil contracts and develops gaps due to the absence of moisture.

11. _____   Vegetables in grocery stores are expensive and in short supply after several months of a widely publicized drought.

12. _____   Reservoirs in California drop to 30 percent of capacity in April.

Exercise 25.2 Precipitation Deficits and Drought

The table below lists the normal seasonal precipitation amounts (inches) at four cities in the United States.

	Dec-Feb (winter)	Mar-May (spring)	Jun-Aug (summer)	Sep-Nov (autumn)	Year
(A) Sacramento	9.2	4.4	0.1	2.5	16.2
(B) Denver	1.9	6.0	4.5	3.1	15.5
(C) St. Louis	6.1	10.7	10.6	8.5	35.9
(D) Washington D.C.	8.3	10.5	11.6	10.8	41.2

1. Draw a line graph showing the seasonal cycle of precipitation at the four cities. Label the line graph for each city as A, B, C, or D.

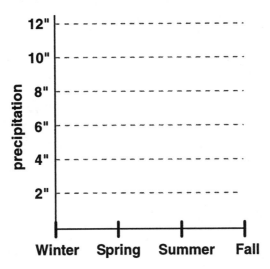

2. Which city would have the largest deficiency of precipitation (in inches) if a drought resulted in a 50 percent deficiency of precipitation in each season? Also, indicate the amount (inches) of the largest deficiency in each season.

    a. Winter (Dec-Feb) _____   b. Summer (Mar-May) _____
    c. Spring (Jun-Aug) _____   d. Fall (Sep-Nov) _____

3. Which city would lose the highest portion of its average annual precipitation if there were a 50% deficiency of precipitation in one season? (Indicate the % in each case).

    a. Winter (Dec-Feb) _____   b. Summer (Mar-May) _____
    c. Spring (Jun-Aug) _____   d. Fall (Sep-Nov) _____

4. List two other factors that should also be considered in assessing the hydrological consequences of a deficit of precipitation?

Exercise 25.3 Understanding Drought

Label each of the following as T (True) or F (False).

1. ____ The mechanisms that initiate drought have been well established by meteorological research.

2. ____ "Rain follows the plow."

3. ____ The wintertime pattern of the jetstream is an important determinant of drought in the western united States.

4. ____ Drought in the central United States is favored by a westward migration of the Bermuda high to a location near the Southeast Coast.

5. ____ Soil moisture plays a role in the ability of a drought to perpetuate or "feed upon itself."

6. ____ During a drought, days can be hotter than normal while nights are cooler than normal.

7. ____ During a drought in the central United States, the jetstream is north of its normal position.

8. ____ The central United States has never had a drought worse than that of the 1930s.

9. ____ Extreme heat always accompanies a drought.

10. ____ A closed high pressure center at 500 mb always precedes a drought in the United States.

11. ____ The northeastern United States is immune to drought because of its proximity to the Gulf Stream.

12. ____ There is no evidence that the settlement of the Great Plains by the European settlers had any impact on the intensity of meteorological drought.

13. ____ El Niño is a robust predictor of drought in the central United States.

Exercise 25.4 Drought Weather Patterns Aloft

On each of the three maps below, draw a 700 mb height pattern that would eventually lead to drought conditions in the indicated region if the pattern became a climatological feature for several months.

DROUGHT IN MID-ATLANTIC STATES

DROUGHT IN CENTRAL U.S.

DROUGHT IN WESTERN MOUNTAIN STATES

Name: _____

Section: _____   Date: _____

Exercise 26.1 Heat Index Calculations

The surface reports below were obtained from stations affected by a summer heat wave. In each case, estimate the Heat Index using the table provided. Additionally, indicate whether the wind will have a warming effect or a cooling effect on a person wearing light clothing.

Temperature	Relative Humidity	Wind	Heat Index	Effect of wind (warming/cooling/none)
90°F	55%	10 mph	_____	_____
103°F	45%	20 mph	_____	_____
86°F	75%	calm	_____	_____
80°F	45%	5 mph	_____	_____
95°F	70%	15 mph	_____	_____

Heat Index as a function of temperature and relative humidity

Relative Humidity (%)

		40	45	50	55	60	65	70	75	80	85	90	95	100
	110	138												
	108	130	137											
	106	124	130	137										
Air	**104**	119	124	131	137									
Temp.	**102**	114	119	124	130	137								
(°F)	**100**	109	114	118	124	129	130							
	98	105	109	113	117	123	128	134						
	96	101	104	108	112	116	121	126	132					
	94	97	100	102	106	110	114	119	124	129	136			
	92	94	96	99	101	105	108	112	116	121	126	131		
	90	91	93	95	97	100	103	106	109	113	117	122	127	132
	88	88	89	91	93	95	98	100	103	106	110	113	117	121
	86	85	87	88	89	91	93	95	97	100	102	105	108	112
	84	83	84	85	86	88	89	90	92	94	96	98	100	103
	82	81	82	83	84	84	95	86	88	89	90	91	93	95
	80	80	80	81	81	82	82	84	84	84	85	86	86	87

Name: _____

Section: _____   Date: _____

Exercise 26.2 Inversions and Surface Heating

Inversions are very stable layers of air. By preventing the mixing of humid air near the surface and drier air aloft, inversions can enhance the build-up of humidity and heat indices near the surface.

For each of the soundings shown below, fill in the blanks for two situations:

	SOUNDING A	**SOUNDING B**
(a) Surface air temperature	_____	_____
(b) height at base of inversion	_____	_____
(c) height at top of inversion	_____	_____
(d) temperature at top of inversion	_____	_____
(e) temperature at base of inversion	_____	_____
(f) magnitude of inversion $(T_{top} - T_{bottom})$	_____	_____
(g) surface air temperature required to eliminate inversion	_____	_____

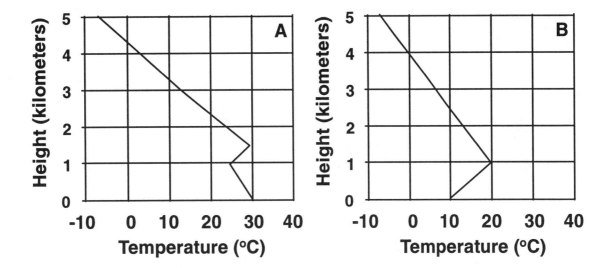

100

Name: _____

Section: _____   Date: _____

Exercise 26.3 Meteorology of Heat Waves

Circle the correct answers to indicate how each of the factors below affects a heat wave and the various measures of heat severity.

1. Evaporation from a wet surface will (increase; decrease; not affect) the Heat Index.

2. Cloud cover will (increase; decrease; not affect) the apparent temperature.

3. Northward migration of the jetstream will (increase; decrease; not affect) the likelihood of a heat wave.

4. An intensifying surface high 500 miles to the west of a location will (increase; decrease; not affect) the likelihood of a heat wave at that location.

5. Westerly winds along the East Coast during the summer (increase; decrease; do not affect) the likelihood of an East Coast heat wave.

6. Westerly winds along the West Coast during the summer (increase; decrease; do not affect) the likelihood of a West Coast heat wave.

7. A large or "steep" environmental lapse rate favors (high; low) humidities at the surface during a heat wave.

8. Nighttime minimum temperatures during a heat wave can be expected to be highest when the overlying airmass is (continental Polar; continental Tropical).

9. Surface high pressure centers are most often found to the (east; west) of areas affected by heat waves.

10. A persistent heat wave is often associated with a closed (high; low) pressure center at the 700 mb level.

11. The vertical motions in the middle troposphere above an area affected by a heat wave are generally (upward; downward).

12. The drier the ground, the (more; less) solar energy is used for heating the ground and the air near the ground.

13. During a heat wave, the temperature difference between urban and rural areas is generally greatest (late at night; late in the afternoon).

Name: _____

Section: _____   Date: _____

Exercise 26.4 Heat Wave Weather Patterns

Two maps are provided below. On the top map, draw the sea level pressure pattern that is commonly associated with a heat wave on the central Plains of the United States. On the bottom map, draw the sea level pressure pattern for conditions that might lead to a heat wave in California (Hint: downslope winds off the Sierra Nevada are common during California heat waves).

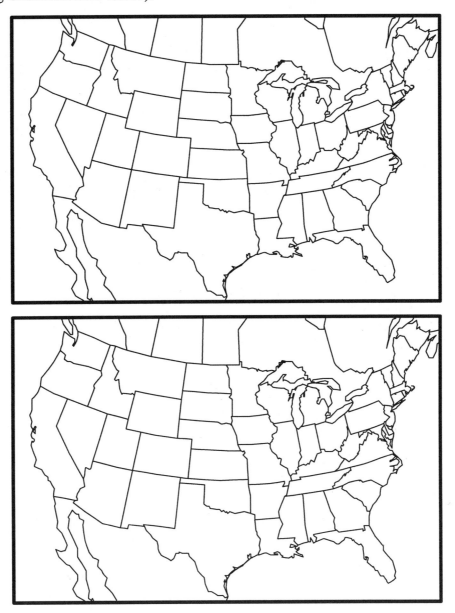